Frontiers in Horticulture

(Volume 1)

(Medicinal and Aromatic Plants: The Basics of Industrial Application)

Edited by

M. Paz Arraiza
Department of Forestry and Environmental Engineering and Management, Technical University of Madrid, Madrid, Spain

A. González-Coloma
Institute of Agricultural Sciences. Department of Crop Protection Serrano 115, 28006 - Madrid, Spain

J. Burillo
Agrifood Research and Technology Centre of Aragon, Department Medicinal and Aromatic Plants, Spain

&

C. Calderón-Guerrero
Department of Forestry and Environmental Engineering and Management, Technical University of Madrid, Madrid, Spain

Frontiers in Horticulture Vol. 1

Volume # 1

Medicinal and Aromatic Plants: The Basics of Industrial Application

Editors: M. Paz Arraiza, A. González-Coloma, J. Burillo and C. Calderón-Guerrero

ISSN (Online): 2588-7874

ISSN (Print): 2588-7882

ISBN (Online): 978-1-68108-550-0

ISBN (Print): 978-1-68108-551-7

First published in 2017.

General:

1. Any dispute or claim arising out of or in connection with this License Agreement or the Work (including non-contractual disputes or claims) will be governed by and construed in accordance with the laws of the U.A.E. as applied in the Emirate of Dubai. Each party agrees that the courts of the Emirate of Dubai shall have exclusive jurisdiction to settle any dispute or claim arising out of or in connection with this License Agreement or the Work (including non-contractual disputes or claims).
2. Your rights under this License Agreement will automatically terminate without notice and without the need for a court order if at any point you breach any terms of this License Agreement. In no event will any delay or failure by Bentham Science Publishers in enforcing your compliance with this License Agreement constitute a waiver of any of its rights.
3. You acknowledge that you have read this License Agreement, and agree to be bound by its terms and conditions. To the extent that any other terms and conditions presented on any website of Bentham Science Publishers conflict with, or are inconsistent with, the terms and conditions set out in this License Agreement, you acknowledge that the terms and conditions set out in this License Agreement shall prevail.

Bentham Science Publishers Ltd.
Executive Suite Y - 2
PO Box 7917, Saif Zone
Sharjah, U.A.E.
Email: subscriptions@benthamscience.org

BENTHAM
SCIENCE

CONTENTS

FOREWORD

Medicinal and Aromatic Plants (MAPs) have been accompanying mankind from its very early beginnings, still mainly due to their tremendous diversity even today these plants constitute a relatively less explored group of economic plant systems.

MAP diversity implies both practical and scientific aspects. The initial practical aspects refer to the traditional uses of plants, mostly in healing and medication. According to a frequently cited data in the special literature, approx. 80% of the world's population makes use of plants or plant derived substances in their every-day medication (Akerele, 1992). In order to curb the overexploitation of these valuable natural resources, since the Chiang Mai Declaration of 1988, much attention has been paid to their sustainable use and exploitation. In addition to sustainable wild-crafting from the natural populations, the cultivation of MAPs is being regarded as an important, sustainable form of dependable raw material production of good quality crude drugs to be used safely and with efficacy in an extremely wide range of uses. Precondition for all this is a sound knowledge of the influencing factors and the targeted transfer into practice of scientific, as well as practical knowledge.

It is the above-mentioned knowledge that is summarized in present eBook, in a relatively concise though transparent way. The volume is divided into two major Parts, Part I Medicinal and Aromatic Plants growing and uses and Part II A study case of MAPs production, uses and commercialization. *Artemisia absinthium* var. Candial.

Starting out from the botanical and chemical aspects of MAPs (active principles incl. identification, isolation and purification) through basic knowledge about most frequently applied methods of horticultural production technology, harvesting, as well as processing Part I. is a virtual storehouse of valuable information. Importantly, insight is provided into the achievements of a wide range of sciences involved. In harmony with the modern requirements on safe medicinal plant use, the subsequent chapters deal with quality management, quality control of MAPs.

As a quasi evaluation of all the knowledge on the uses and commercialization of MAPs discussed in the previous chapters, the book offers Readers a case study on wormwood (*Artemisia absinthium* var. Candial), a woody perennial species of the genus Artemisia which is native mostly to the temperate regions of Eurasia and Northern Africa and is widely naturalized also on the North American continent. The selection of wormwood as a model plant can be regarded symbolic, since this species is known and used from the early times of human civilizations. As such, it can be regarded as an example for the timelessness diversity and wealth of medicinal and aromatic plants. In addition, wormwood is grown not only as an ornamental plant but it has acquired reputation as an ingredient of the famous European beverage absinthe. Its manifold uses include expelling round worms and threadworms, treating irritations, sprains and bruises. Planted as a companion plant it is reputed to repel slugs, snails, aphids, etc. In cultivation it can be propagated both vegetatively by semi-hardwood cuttings and by seeds, so that this ancient plant constitutes a good example to illustrate many of the principles related to the cultivation, processing and utilization of medicinal and aromatic plants described in the previous chapters of Part 1. It is a special value of the book that the Authors collective of well-known representatives of their profession have paid a special attention to the requirements of quality management including the characterization and valorization of extracts obtained from wormwood which in the case of all MAPs is of great importance.

This e-book is expected to be a useful addition to the not too abundant science based special literature on MAPs. It is a multi-faceted volume with a wealth of useful and up-to-date information for those who wish to orientate themselves in the science and practice of medicinal aromatic plant diversity.

Habil. Prof. Dr. MÁTHÉ
President of International Council for Medicinal and Aromatic Plants
Hungary

PREFACE

Nature has been a source of medicinal agents for thousands of years, since healing with the plants dates back probably to the evolution of Homo sapiens. Medicinal and Aromatic Plants (MAPs) produce a range of secondary metabolites useful for humans, making this group a target of scientific interest and market. These products are substances ranging from spices and condiments, to essential oils and extracts, used in cosmetics, perfumery, food, nutraceuticals, natural dyes, natural biocides and many other compounds.

The demand for these plants and the products obtained from them is nowadays increasing worldwide, and some of them are becoming scarce in some areas due to wild harvesting. The demand in Europe for products obtained from MAPs is going to grow, for our health care needs and demands are changing. In fact, medicinal use of MAPs represent the biggest number of plant species used by human, as they provide most of the medicinal ingredients in many health care traditional systems. But these plants are not only used for medicines. Their use in the cosmetic and perfumery, food and liquor industries is well known, as well as other uses as natural pesticides, dyes, tannins, etc. These other uses of MAPs are many important for their growing economic importance and market possibilities. Plant extracts have been used as insecticides since before ancient Rome, a practice that continues today with more than 2000 species.

Currently, there is considerable decentralization of this market, which is characterized by lack of information and data in the sector, together with a deficiency regarding quality requirements of crops and products obtained from them, method of production and cultivation, market demand and how to access to it. At the same time, in many countries, wild collection is or has been practiced. This activity often lacks of quality control mechanisms of the raw material generating environmental degradation.

MAPs are a source of biodiversity, local development, and entrepreneurship, preventing rural population emigrating to urban areas, and a potential for development of ecotourism businesses. This field is a niche for market, business and entrepreneurship, demanding the knowledge and tools required for a quality product and the necessary skills for its start-up. They represent a relatively new area of social environmental education with considerable and growing interest.

This eBook focuses on establishing a fundamental understanding of the tradition and science that envelops MAPs. It aims to provide a basis on horticulture, active principles, identification, isolation and purification as well as the handling, processing and uses of MAPs and their derivatives.

The book is made up of two parts. A general first part, starting with a general vision describing the most important topics dealing with the production and processing of MAPS, concepts and sources of active principles, followed by cultivation an production of MAPs, what are active ingredients and their classification, the obtaining of raw matter, and the uses of these plants in the cosmetic, food, medicinal and other sectors, MAPs sustainable assessment and economic valuation of environmental services provided by MAPs.

The second part is a practical case where the plant *Arthemisia absinthium* var. Candial is assessed in terms of propagation, production and collection. For many years, the authors have worked on this specie, domesticated it and identified its active principles and their uses, presenting a unique research work that encompasses al the aspect s presented in the general part, applied to a single species of high interest. This part gives the book its unique and specific value in this area of interest. This case study is what gives the novel and unique character of the book, which will certainly be of great demand.

This book is intended for students and professionals in agricultural engineering, pharmacy, forestry, chemistry of natural products, research and education institutions, libraries. Also, it is focussed on general public interested in aromatic and medicinal plants production, transformation and commercialization. This book is a reference book that brings a new vision, a concrete example that has been studied in depth at all levels covered in the book.

ACKNOWLEDGEMENTS

I want to dedicate the book to the memory of two very good friends who contributed their bit to the world of aromatic and medicinal plants and that left their trajectory before being able to complete it. Pedro Manuel Díaz Fernández, my neighbour, career partner and colleague, but above all, my friend, great expert in wild edible plants, whom I would have invited to write a chapter in this book. Andrés Corral, an entrepreneur, fighter and friend, who with his strength and spirit reached the highest position in essential oil production for perfume industry. The seeds that you both planted will continue to grow and the fields in bloom will bring us their aromas in your memory.

M. Paz Arraiza
PhD. Associate Professor
Department of Forest Engineering,
ETSI Montes, Forestal y Medio Natural,
Universidad Politécnica de Madrid,
Ciudad Universitaria SN. 28040 Madrid,
Spain
Tel: 0034 91 336 64 08
E-mail: paz.arraiza@upm.es

List of Contributors

A. González-Coloma	Instituto de Productos Naturales y Agrobiología, CSIC, Tenerife, Spain
C. Calderón-Guerrero	Department of Forestry and Environmental Engineering and Management, Technical University of Madrid, Madrid, Spain
C. E. Díaz	Instituto de Productos Naturales y Agrobiología, CSIC, Tenerife, Spain
D. A. Meloni	Faculty of Agronomy and Agroindustries, National University of Santiago, Santiago, Argentina
J. Burillo	Centro de Investigación y Tecnología Agroalimentaria de Aragón, Gobierno de Aragón, Departamento de Ciencia, Tecnología y Universidad,, Zaragoza, Spain
J. N. Rocha	Centro de Investigación y Tecnología Agroalimentaria de Aragón, Gobierno de Aragón, Departamento de Ciencia, Tecnología y Universidad, , Zaragoza, Spain
J. Tapia	Facultad de Economía y Empresa, Departamento de Económico, Universidad de Zaragoza, Zaragoza, Spain
J. V. López	Department of Forestry and Environmental Engineering and Management, Technical University of Madrid, Madrid, Spain
L. F. Julio	Instituto de Ciencias Agrarias, CSIC, Madrid, Spain
M. A. Sarmiento	Faculty of Forestry, National University of Santiago del Estero, Santiago del Estero, Argentina
M. F. Andres	Instituto de Ciencias Agrarias, CSIC, Madrid, Spain
M. P. Arraiza	Department of Forestry and Environmental Engineering and Management, Technical University of Madrid, Madrid, Spain
S. C. Guillén	Ecology Department, Agronomy Faculty, National University of Tucuman, Tucuman, Argentina

CHAPTER 1

Introduction: Identification of Potential MAPS: Ethnobotany as a Source of Active Principles

M. Paz Arraiza[1,*], **Carlos Calderón-Guerrero**[1], **José V. López**[1], **Silvia C. Guillén**[2], **Miguel A. Sarmiento**[3] and **Diego A. Meloni**[4]

[1] *Department of Forestry and Environmental Engineering and Management, Technical University of Madrid, Spain*

[2] *Ecology Department, Faculty of Agronomy, National University of Tucuman, Argentina*

[3] *Faculty of Forestry, National University of Santiago del Estero, Argentina*

[4] *Faculty of Agronomy and Agroindustries, National University of Santiago del Estero, Argentina*

Abstract: Plants produce useful substances for human health and care. These compounds are usually secondary metabolites, which do not have a direct function in the development and growth of the plant, but help the plant interacting with its environment. They are produced by certain tissues and cells, during specific stages of the plants or at different moment of their living cycle. They act as insects attractive (for example for pollination) or repellent agents, as a defence from extreme weather conditions (drought, freezing), to avoid growth of unwelcome plants (allelopathy), or to combat bacteria and viruses. The useful part is called plant drug, and it can be the whole plant or a part of it. These active principles can be extracted by distillation, solvent extraction, pressure or other methods, depending on to their chemical characteristics. The demand for these plants and the related products is nowadays increasing worldwide, and some of them are becoming scarce in some areas due to wild harvesting. In Europe, the demand for instance for food supplements, is going to grow, for our health care needs and demands are changing. In fact, MAPs for medicinal use represent the highest number of plant species used by humans, as they provide most of the medicinal ingredients in many health care traditional systems. But these plants are not only used for medicines. Their use in the cosmetic and perfumery, food and liquor industries is well known, as well as other uses as natural pesticides, dyes, tannins, *etc.*

Keywords: Active Principles, Aromatic Plants, Ethnobotany, Health care, Herbal medicine, History, MAPs, Medicinal Plants, Medicinal use, Natural pesticides, Wild harvesting.

* **Corresponding author María Paz Arraiza:** Department of Forestry and Environmental Engineering and Management, Technical University of Madrid, Madrid, Spain; Tel/Fax: +34 91 336 64 08; Email: paz.arraiza@upm.es

M. Paz Arraiza, A. González-Coloma, J. Burillo and C. Calderón-Guerrero (Eds.)

INTRODUCTION

Plants produce useful substances for human health and care. These compounds are usually secondary metabolites, which do not have a direct function in the development and growth of the plant, but they help the plant interact with its environment. They are produced by certain tissues and cells, during specific stages of the plants or at different moment of their living cycle. They act as insects attractive (for example for pollination) or repellent agents, as a defence from extreme weather conditions (drought, freezing), to avoid growth of unwelcome plants (allelopathy), or to combat bacteria and viruses.

The useful part is called plant drug, and it can be the whole plant or a part of it. These active principles can be extracted by distillation, solvent extraction, pressure or other methods, depending on to their chemical characteristics.

The demand for these plants and the products obtained from them is nowadays increasing worldwide, and some of them are becoming scarce in some areas due to wild harvesting. According to CBI [1], (Centre for the Promotion of Imports from Developing Countries), the demand in Europe for products obtained from MAPs, for instance food supplements, is going to grow, for our health care needs and demands are changing. In fact, MAPs for medicinal use represent the biggest number of plant species used by humans, as they provide most of the medicinal ingredients in health care traditional systems [2].

These plants, however, are not only used for medicines. Their use in the cosmetic and perfumery, food and liquor industries is well known, as well as other uses as natural pesticides, dyes, tannins, *etc.*

These plants, the active principles they produce, the extraction, isolation and identification methods, cultivation and different industrial uses, will be discussed through the chapters of this book.

Basic Concepts [3]

Active Ingredients are the plant components providing therapeutic activity. When these active ingredients are identified, the medicinal product must be standardized, so that the concentration of the active ingredient in it is known. When the identification of the active principle(s) is not possible, the whole herbal medicine is considered the active ingredient *per se*.

Aromatic Plants are medicinal plants containing essential oils.

Condiments or Spices are used for their smell, fragrance and taste characters, which give food and drinks aromas, colours and flavours that make them more appetizing, tasty and pleasing to smell, sight and taste.

Dye Plants contain substances that are employed to dye vegetal or animal fibres and fabric (wool, leather, cotton, linen).

Essential Oils are the volatile organic compounds present in the plant material. They are usually extracted by distillation, although sometimes expression is employed, such as in the case of citrus fruits.

Extracts are the non-volatile organic fraction present in the plant material. They are usually extracted by solvent extraction.

Finished Herbal Products are made from one or more plants, in such case they are called mixed herbal product. Finished herbal products and mixed herbal products may contain excipients, but if chemical active ingredients are added, they are no longer considered herbal products.

Herbal Materials include also juices, gums, oils, essential oils, resins and plant powders. They can be processed by several local procedures, for example steaming, roasting, baking, or mixed with alcohol to produce alcoholic drinks, among other treatments. The specific useful part of the plant, which provides the medicinal properties, is called **plant drug**.

Herbal Preparations are the raw materials for other herbal products, like powders, extracts, tinctures and oils. They are obtained by distillation, extraction, fractionation, purification, concentration, among other processes.

Herbs are crude plant material. This term includes leaves, flowers, fruit, seed, stems, wood, bark, roots, rhizomes or other parts of the plant, entire, fragmented or powdered.

Oleoresins are obtained from spices such as paprika by solvent extraction followed by solvent evaporation at low temperature and partial vacuum. They are liquid or semisolid substances containing the aroma and taste of the spice (including the no volatile components).

Therapeutic Activity indicates the capacity of these products for preventing and treating illnesses; to improve or eliminate their symptoms, and to improve health at different levels.

Traditional use of Herbal Medicines takes into account the historical use of these products, which establishes their use, known to be safe and effective. Their use is usually accepted by authorities.

Main Uses of MAPs in Industry

The main industrial uses of MAPs and of their essential oils are cosmetic, food processing, and medicinal industries. According to European vegetal product turnover [4], Germany for instance is by far the largest market in natural cosmetics market with 8% market share.

Essential oils are gaining importance in flavouring indoor environments in Japan, where different essences are used depending on the desired emotional state. Aromatherapy, also accepted in hospitals in some countries, is increasing the demand for essential oils of high quality. It is also increasing their use as antiseptic and their effect against some viruses as in the case of eucalyptus, tea tree and thyme oil [5].

Seasonings or spices are used as natural additives and antioxidants in the food manufacturing industry. Oregano, rosemary and sage are used in meat industries as preservatives. They avoid rancidness of products without the need of adding chemical synthetic antioxidants, preservatives and stabilizers.

Food supplements do not need special licenses and there are not restrictions to their use. The preference for natural foods has led to replace artificial colours and flavours, promoting the natural herbs. The boom in the microwave, frozen and fast food with new tastes demands the use of more spices and herbs. The candy and cosmetics multinationals also demand all types of essential oils and aromas.

HISTORY OF AROMATIC AND MEDICINAL PLANTS

Medicinal ad Aromatic Plants (from now on, MAPs) history is as long as human history.

The use of medicinal plants is so remote that the distillation has been practiced for thousands of years in China and India. Traditional Chinese medicine has been used from ancient times, being botanicals their main source of remedies [6]. In India, Ayurveda medical system has been practised for nearly 5000 years [7].

Undoubtedly, one of the largest contributors was the Egyptian civilization. Moreover, in The Ebers Papyrum (2278 BC.) and Smith Papyrum (2263 BC) they cited how to prepare and cultivate a number of drugs: opium (*Papaver somniferum*) and hemp (*Cannabis sativa*) [8]. The importance and implication of ancient Egypt in this matter was so high that the first of his notable physicians, Imhotep, was promoted to the rank of God thanks to the miraculous cures attributed to him.

Traditional Chinese medicine has been used since ancient times. The main source of Chinese medicines is botanical, though animal and mineral materials are used as well. Traditional medicine is still in practice in China nowadays. Most of the population still uses traditional drugs, mostly in rural areas. There are more than 5000 traditional remedies, accounting one fifth of the Chinese pharmaceutical market [9]. The most famous book on plant medicines is the Pen Tsao, written by the Emperor Shen Nung (2500 BC), where 365 drugs are listed, among which camphor (*Cinnamomum camphora*), tea (*Camellia sinensis*), Mayapple (*Podophyllum peltatum*), yellow gentian (*Gentiana lutea*), ginseng (*Zingiber officinale*), jimson weed (*Datura stramonium*), cinnamon bark (*Cinnamomum zeylanicum*), and ephedra (*Ephedra sinensis*) [10]. The specific name *sinensis* make reference to plants native to China.

In India, the traditional medicines mostly come from the Ayurveda, which is a medical system practised in India for more or less 5000 years. It considers the body as a whole, and employs diet and herbal medicines for the prevention of illness, of the body, mind and spirit. In India, traditional systems have remained quite separate from Western medicine. Plants coming from Indian Ayurvedic medicine are for example snake root (*Rawolfia serpentina*), nutmeg (*Myristica fragans*), pepper (*Piper nigrum*) and clove (*Eugenia caryophillata*) [11].

In Europe the history of MAPs starts with the Early Greeks and Romans. Many Greek and Roman scientists recorded the use of plants.

Hippocrates (Kos, Greece, 470 – 377 BC) was one of the most important physicians and the father of medicine. He wrote “Corpus Hippocraticum”, a compendium of books with the medical knowledge of Kos. Hippocrates introduced the concept of ‘physis’ and changed from hieratic or theocratic medicine into rational medicine, separating disease from religion or beliefs and rationalizing the origin of disease in food, habits and environmental facts [12].

Dioscorides (Anarzarbos, Greece, 40 to 90 AD) described more than 600 plants remedies documented in the book “Materia Medica”. It was the quintessential treatise on medicine that was read, copied and edited for over 1500 years until it was supplanted by the herbaria in the Renaissance [13].

Galen (Pergamo, Turkey, 129 AD) was a disciple of Hippocrates, and the use of drugs on a large scale is attributed to him. He gathered healing plants and prepared prescriptions based on complex mixtures. The word “Galenic” in its modern sense refers to the science of preparing medicines.

The Arabs used many plants that they brought to Europe after their expansion in the VII – VIII centuries. Many of them came from India, a country with which they had commercial relationships. They used aloe (*Aloe vera*), deadly nightshade (*Atropa belladonna*), henbane (*Hyoscyamus niger*), coffee (*Coffea arabica*), ginger (*Zingiber officinale*), strychnos (*Strychnos nux-vomica*), saffron (*Crocus sativus*), curcuma (*Curcuma longa*), pepper (*Piper nigrum*), cinnamon (*Cinnamomum verum*), rhubarb (*Rheum rhabarbarum*) and senna (*Cassia senna)*, among others [10].

Persian medicine's greatest representative is Abu Ibn Sibna, (Afshana, Uzbekistan, 980), whose name is better known by its Latinized form Avicenna. He was the most famous Persian physician, wrote, in 1012, the Canon of Avicenna, a compendium of all existing medical knowledge at the time. The Canon consisted of 5 specific books, on general knowledge of health, pharmacology, pathology, symptoms and diagnosis. Rhazes (865–925 AD) and Avicenna (980–1037 AD) are accredited as the founders of the golden age of Persian medical sciences that took place from the eighth to seventeenth centuries. Most current ethnopharmacological knowledge in Iran has been derived from historical manuscripts known as *Qarabadin*, meaning pharmacopoeia, which consist of medical texts on drug compounds, formulas, indications and pharmaceutical applications. They list numerous forms of medicinal oils prepared from herbs, known as *Dohn* (*Adhaan* in plural). Examples of medicinal herbs in Persian traditional medicine include castor oil (*Ricinus communis*), hemp (*Cannabis sativa*), labdanum (*Cistus ladanifer*), olive oil (*Olea europaea*) and saffron (*Crocus sativus*) [14].

Paracelsus (Einsiedeln, Switzerland, 1493 - 1541). Swiss physician and chemist, who ignored the legacy of Galen and Avicenna and focussed medical treatments on the free action of natural processes. He also published a book on surgery. His contributions include the first clinical description of syphilis, and new treatments based on mineral substances such as lead or mercury.

Carl Linnaeus (Uppsala, 1707 – 1778) was a Swedish doctor in medicine (he was interested in the study of botany, but at that time that was only possible by studying medicine). In 1749 he published *Materia Medica*, a book reviewing plants, minerals and animals used in medicine, such as opium poppy (*Papaver somniferum*), valerian (*Valeriana officinalis*), deadly nightshade (*Atropa belladonna*), sweet wormwood (*Artemisia annua*) and chamomile (*Matricaria recutita*), which gave place to the first Swedish Pharmacopoeia in 1775. He is the father of the scientific binomial nomenclature of living beings.

Medicine during the middle ages was a mixture of existing ideas from antiquity and spiritual influences [15]. It was thought that people had four humours linked to the four elements (fire, wind, earth and water) and techniques such as bleeding were used to treat illnesses. Standard medical knowledge was based upon Greek and Roman texts preserved in monasteries. Monasteries developed herb gardens for medicine and wrote herbal books and manuscripts. Dioscordes' *De Materia Medica* was widely copied and translated at that time. Medicinal plants used included sage (*Salvia officinalis*), angelica (*Angelica archangelica*), primrose (*Primula vera*), lungwort (*Pulmonaria officinalis*), coltsfoot, thyme (*Thymus vulgaris*), rue (*Ruta graveolens*) among many.

After these dark ages, the Renaissance supposed a change, with the revival of herbalism. Medicinal plants were identified, and this, coupled with the printing press, started the so called the "Age of Herbals". Herbals were treaties on medicinal plants, their uses, and properties, handsomely decorated.

For instance, the famous Doctrine of Signatures stated that the medicinal use of a plant could be established by the identification of the morphological characteristics of the plant with the disease it cured.

Thus, the red juice of bloodwort (genus *Sanguinaria,* Papaveraceae) indicated the plant's ability to cure blood disorders; the lobed liver like morphology of liverworts (division *Marchantiophyta*) made them it useful for treating liver illnesses; the human-like shape of mandrake root (*Mandragora officinarum,* Solanaceae) made it useful for enhancing virility and promoting conception; the brain morphology of walnuts suggested their use for treating mental disorders.

When the Europeans reached America, they learned the local knowledge on medicinal use of plants. The Aztecs used more than 3000 medicinal plant species. The first book on Mexican medicine, *Medicinalibus Indorum Herbis*, was written in 1552 by an Aztec physician, Martín de la Cruz, and later translated into Latin by Juan Bernardino [16]. Among the species that were brought to Europe from the Americas, it is possible to list quinine (*Cinchona* sp, Rubiaceae), coca (*Erythroxylum coca,* Erythroxylaceae) or tobacco (*Nicotiana tabacum,* Solanaceae).

Herbal Medicine Today

The most important step for incorporating drug plants to modern medicine started with the isolation, in the 19th and early 20th centuries, of the active molecules present in them: first morphine from opium, followed by cocaine, codeine, digitoxin and quinine. Isolation was the starting point of modern medicine. This started a research tendency to identify medicinal plant composition, that lead to the beginning of pharmacologic, perfumery and seasoning industry.

At that time, first synthetic drugs based on natural plants started to be formulated. One example is salicylic acid, an active ingredient known as pain reliever. It was firstly synthesized in 1853, leading to the development of aspirin.

Synthetic molecules started then to be the main source of medicines. While in the 20th century Western medicine stayed away from herbalism, 75% to 90% of the rural population of the rest of the world still relied (and relies today) on herbal medicine as their only health care resource.

In the 21st century, in modern developed countries there is a renewed interest for the use of MAPs in the pharmacological, nutrition, perfumery and cosmetic fields. The demand for a healthier lifestyle and a higher standard of living, promoted the research for natural products, but not losing quality and properties. In countries like Germany, the consumption of "natural" medicines accounts 30-40% of medical prescriptions [17]. The offer of products of natural origin is increasingly up.

Nowadays, in western countries, natural medicine complements traditional medicine, and essential oils and products with a medicinal plant origin are everyday more demanded.

In this moment, the big suppliers of medicinal plant are Eastern European countries (Romania, Bulgaria), Turkey and Far East (India, Hong Kong and China) [18]. Enormous quantities of raw matter are being produced, and in many cases, quality is less important than quantity, as a result of no environmental and quality control regulations.

China is the example of a country which incorporates traditional herbal medicine in its modern health care system, which mixes herbal medicine, acupuncture, and Western medicine. They grow thousands of medicinal plant species making them available for the pharmaceutical markets. Prescriptions are usually made of measured amounts of specific herbs rather than synthetic medicines.

Spain, with a high medicinal herb biodiversity, has been a supplier of high quality material commonly from wild harvesting, nowadays regulated. The quality production of medicinal plant is a good alternative for little supplier in traditional agriculture, currently dealing with serious economic problem. The development of the medicinal plant production requires a trade regulation framework that has been delayed by the lack of understanding between pharmacist and herbalist industries.

Medicinal plants are also vulnerable of extinction through wild harvesting. More than 150 European plant species are threatened especially in Albania, Hungary,

Spain and Turkey by collection and overuse [19]. According to these authors, the most threatened species are Pheasant's eye (*Adonis vernalis),* Bearberry (*Arctostaphylos uva-ursi),* Arnica (*Arnica montana),* Island moss (*Cetraria islandica),* Sundew (*Drosera rotundifolia),* Yellow gentian (*Gentiana lutea),* Liquorice (*Glycyrrhiza glabra);* Gypsophila (*Ankyropetalum gypsophiloides*), Bogbean (*Menyanthes trifoliata)*, Butcher's broom (*Ruscus aculeatus)*, some species of genus *Sideritis* spp., and species of genus thyme and oregano (*Thymus* spp., *Origanum* spp., and *Thymbra* spp.).

Ethnobotany as a Source of Active Principles

Nature has been a source of medicinal products for millennia, with many useful drugs developed from plant sources. Here is a small list of some drugs derived from plants traditionally used in some cultures (Table **1**) [20].

Table 1. Some active compounds derived from plants of traditional use.

Uses	Active Compound	Plant
Malaria	Quinine	*Cinchona* spp
	Artemisine	*Artemisia annua*
Hypertension	Reserpine	*Rawolfia serpentina*
Anticancer	Etoposide and tenoposide	*Catharantus roseus*
	Paclitaxel, Taxol ®	*Taxus brevifolia*
Migraine	Ergotamine	*Claviceps purpurea*
Heart diseases	Digitalin	*Digitalis purpurea*
Analgesic	Codeine, Morphine	*Papaver somniferum*
Anesthesic	Atropine	*Atropa belladonna*
Glaucoma	Pilocarpine	*Pilocarpus* spp

The research on the use of plants as drugs is a multidisciplinary process which includes the documentation of the traditional knowledge, the quantitative determinations of plants used (importance, distribution, abundance *etc.*), the phytochemical, pharmacological and toxicological evaluation and the development of mechanisms for community compensations [21].

For decades, scientists in this field have focused their work towards:

- The discovery of new functional compounds.
- The identification of their metabolic pathways.
- The identification of the genes for the metabolic pathways.
- Their production in cell culture.

- The study of secondary metabolism at whole plant level.
- The implementation of large cell culture systems and purification of secondary metabolites.

The potential value of tropical plants medicines is considerable and several programs have been developed prospecting to investigate the pharmacological activity of the components of tropical forests, such as [22].

- The Merck-INBio Convention plants in Costa Rica.
- The program search for active compounds against Cancer and AIDS National Cancer Institute in the US.
- International Groups for Cooperation in Biodiversity, (US, 1993), financed by National Institute of Health, National Science Foundation and the Agency for International Development.
- The draft biochemical prospecting Yutajé rainforest in Venezuela.

In order to protect traditional heritage and knowledge when using ethnobotany as a source of MAPs, any project related with drugs search from ethnobotanical sources should include the following aspects [23].

a. Documentation of traditional knowledge and its variation in the specific context;
b. The determination the ecological characteristics of useful species, considering the source systems and ethno ecological strategies of use;
c. The phytochemical, pharmacological and toxicological evaluation of medicinal species identified;
d. The development of mechanisms to compensate communities, through the recognition of intellectual property rights of communities and proposing strategies for conservation of plants and their ecosystems of origin.

Today trials on medicinal products from plants are aimed at diseases such as cancer treatments, Alzheimer disease, Parkinson's disease. According to 1993 Nobel Medicine Prize, Richard J. Roberts [24], 90% of funding for medical research in the world is dedicated to treating lifestyle diseases (insomnia, impotence, obesity) that affect only 10% of the population. The remaining 90% suffers more diseases that cause millions of deaths (malaria, tuberculosis, sleeping sickness and Chagas disease) although the budget for them is 10% of the total.

Some of the drugs approved in the last 25 years are shown in Table **2**.

Table 2. Some of the drugs approved in the last 25 years.

Disease	Active compound	Plant
Alzheimer's disease	Galanthamine	*Galanthus nivalis*
Tirosinemia	Nitisinone	*Callistemon citrinus*
Parkinson	Apomorphine	*Papaver somniferum*
COPD (Chronic lung obstruction)	Tiotropio bromuro	*Atropa belladonna*

Some drugs in essay phase against different cancer types and treatments are, among others [25]:

- Combretastatin A4 (isolated from the South African tree *Combretum caffrum)*, it was referred to combretastatin A4 phosphate and AVE-8062. These analogs bind to tubulin leading to morphological changes disrupting the tumor vasculature.
- Homoharringtonine (an alkaloid from the Chinese tree C*ephalotaxus harringtonia),* an inhibitor of protein synthesis studied for its activity against leukemia.
- Ingenol 3-O-angelate, analogue to polyhydroxyl ingenol diterpenoid, obtained from *Euphorbia peplus*, is a potential topical chemotherapy agent for skin cancer and shows its action through activation of protein kinase.
- A synthetic analogue of daidzein, an isoflavone from soybean (*Glycine max*), is being tested as a treatment for cervical cancer, ovarian, prostate, kidney and vaginal cancers.
- Protopanaxadiol, derived from a triterpene aglycone from ginseng (*Panax ginseng*), has an effect on cancer cells through various signalling pathways, and also against resistant tumours.
- A triepoxide diterpene was isolated from *Tripterygium wilfordii*, and is being used for autoimmune and inflammatory diseases in China, such as rheumatoid arthritis. PG490-88 (14-succinyl triptolide sodium salt), a semisynthetic triptolide analog, exerts antiproliferative prostate epithelial cell activity and tumour regression in colon and lung cancer.

Table **3** shows top ten ranked medicinal plants most commonly used in the United States and Europe [26].

Table 3. Top ten ranked medicinal plants most commonly used in the United States and Europe.

Species	Use
Hypericum perforatum	Anxiety, depression, insomnia
Echinacea purpurea	Immune stimulation

(Table 3) contd.....

Species	Use
Ginkgo biloba	Dementia, Alzheimer's disease, tinnitus
Sabal serrulata	Benign prostatic hyperplasia
Tanacetum parthenium	Migraine prophylaxis
Allium sativum	Lipid-lowering, antithrombotic, fibrinolytic, anti-hypertensive, anti-atherosclerotic
Zingiber officinalis	Antiemetic
Panax ginseng	Tonic, performance enhancer, adaptogen, mood enhancer
Valeriana officinalis	Sedative, hypnotic, anxiolytic
Ephedra distachya	Asthma, rhinitis, common cold, weight loss; enhancer of athletic performance

Identification of Potential MAPs

Targeted selection is usually based on ethnopharmacology (the study of the use of local materials by an ethnic community after several generations of living in their particular local ecosystem).

Balunas and Kingdom [27] propose this strategy for drug development from plant sources, which will lead to an herbal medicine or an isolated active compound. This strategy combines the following criteria:

- Observation of plants of traditional use (ethnobotany or ethnopharmacology).
- Identification of their chemical constituents.
- Assessing its toxicity.
- Randomised selection of plants to be assayed.
- Combination of several of these criteria

The most common strategy would be [28]:

- Observation of the use of natural plants in traditional medicine in different cultures.
- Study of the preparation procedure used by these cultures for adequate extraction methods.
- Formulation used will give information about the pharmacological activity.
- It will be necessary also to translate the signs and symptoms to western medical concepts.

Another strategy proposed by the same authors, and this one is the most used when searching for drugs for cancer treatment is to perform a randomised search based on a pharmacological activity, the result of which will be a selection of species for further study.

Random selection consists of acquiring a large number of samples from a locality or from a collection of plants, and submitting them to a particular biological test. There is not only one criterion for the selection, just to ensure diversity of species and habitats. This approach is taken when the target disease is unknown or not common in traditional medicine. Anticancer treatment (Paclitaxel) was discovered this way [27].

According to S.M.K. Rates [28], in order to prepare the plant and to isolate the active components, once the plant is chosen, the next steps will be followed:

- Botanical identification recording place and date of collection.
- Recollection and stabilisation, this is usually done by air drying at room temperature or oven dry.
- Powering and extraction of the plant material. If the active molecules are known, recommended methods will be used, if not, the method will be selected based in the folk uses or by extraction of solvents of different polarities.
- Sequential fractionation for purification and isolation of active molecules.
- Thin layer chromatography (TLC), gas chromatography (GC) and liquid chromatography (HPLC) will be performed for identification of isolated molecules.
- Spectroscopic methods (UV, IR, MS, NMR) will be performed for identification of molecules.
- Each fraction will be bio assayed an assessed for toxicity using microorganisms, molluscs, insects, cellular systems, cell cultures, isolated organs or *in vitro* tests.

Tests must be simple, reproducible, fast and cheap [29, 30]. The final steps are a total or partial synthesis, preparation of analogues and derived products, modulation of the biological activity and the determination of the structure – activity relationship.

CONFLICT OF INTEREST

The authors declare no conflict of interest, financial or otherwise.

ACKNOWLEDGEMENTS

Declared none.

REFERENCES

[1] CBI. The Market for Natural Ingredients for Pharmaceuticals in the EU. Centre for the Promotion of Imports from Developing Countries 2008.

[2] Hamilton AC. Medicinal plants, conservation and livelihoods. Biodivers Conserv 2004; 13(8): 1477-517.
[http://dx.doi.org/10.1023/B:BIOC.0000021333.23413.42]

[3] World Health Organization. General Guidelines for Methodologies on Research on Traditional Medicine. Geneva: WHO Library Cataloguing-in-Publication Data 2000.

[4] Gruenwald J. The market situation and marketing of herbal medicinal products (HMP) in Europe. World Congress on Medicinal and Aromatic Plants for Human Welfare. 2: L-33.

[5] Astani A, Reichling J, Schnitzler P. Comparative study on the antiviral activity of selected monoterpenes derived from essential oils. Phytother Res 2010; 24(5): 673-9. [http://dx.doi.org/10.1002/ptr.2955] [PMID: 19653195]

[6] Li L. Opportunity and challenge of traditional Chinese medicine in face of the entrance to WTO (World Trade Organization). Chin. Inform. trad. Chin Med 2000; 7: 7-8. [in Chinese].

[7] Morgan K. Medicine of the Gods: Basic Principles of Ayurvedic Medicine 2002. Available at: http://www.compulink.co.uk/∼mandrake/ayurveda.htm

[8] Tyler VE. Herbal medicine: from the past to the future. Public Health Nutr 2000; 3(4A): 447-52. [http://dx.doi.org/10.1017/S1368980000000525] [PMID: 11276292]

[9] IARC. Monographs on the Evaluation of Carcinogenic Risks to Humans Some Traditional Herbal Medicines, Some Mycotoxins, Naphthalene and Styrene. France: IARC Lyon 2002; Vol. 82.

[10] Petrovska BB. Historical review of medicinal plants' usage. Pharmacogn Rev 2012; 6(11): 1-5. [http://dx.doi.org/10.4103/0973-7847.95849] [PMID: 22654398]

[11] Ernst E. Ayurvedic medicines. Pharmacoepidemiol Drug Saf 2002; 11(6): 455-6. [http://dx.doi.org/10.1002/pds.756] [PMID: 12426929]

[12] Orfanos CE. From Hippocrates to modern medicine. J Eur Acad Dermatol Venereol 2007; 21(6): 852-8. [http://dx.doi.org/10.1111/j.1468-3083.2007.02273.x] [PMID: 17567335]

[13] Magner L. A History of Medicine. NY: Marcel Dekker, Inc. 1992.

[14] Hamedi A, Zarshenas MM, Sohrabpour M, Zargaran A. Herbal medicinal oils in traditional Persian medicine. Pharm Biol 2013; 51(9): 1208-18. [http://dx.doi.org/10.3109/13880209.2013.777462] [PMID: 23746335]

[15] Hajar R. The Air of History (Part II) Medicine in the Middle Ages. Heart Views 2012; 13(4): 158-62. [http://dx.doi.org/10.4103/1995-705X.105744] [PMID: 23437419]

[16] Aranda A, Viesca C, Sánchez G, Sánchez G, Ramos M, Sanfilippo J. La materia médica en el Libellus de Medicinalibus Indorum Herbis. Rev Fac Med UNAM 2003; 46(1)

[17] Principe PP. Valuing the biodiversity of medicinal plants. The conservation of medicinal plants 1991; 79-124. [http://dx.doi.org/10.1017/CBO9780511753312.008]

[18] Lange D. Crancker *et al.*Ed. Acta Hort.. Medicinal and aromatic plants: trade, production, and Management of botanical resources. Proc XXVI IHC – Future for medicinal and aromatic plants. 629: 177-97. [http://dx.doi.org/10.17660/ActaHortic.2004.629.25]

[19] Lange D. Europe's medicinal and aromatic plants. A TRAFFIC Network report 1998.

[20] Cragg GM, Newman DJ. Biochimica et Biophysica Acta 2013; 1830: 3670-95.

[21] Bermúdez A, Oliveira-Miranda MA, Velázquez D. La investigación etnobotánica sobre plantas medicinales: una revisión de sus objetivos y enfoques actuales. Interciencia 2005; 30(8): 453-459, 514-516.

[22] Avendaño M. Los productos naturales en la búsqueda de nuevos fármacos. Una visión de conjunto María del Carmen Avendaño López Anales de la Real Academia Nacional de Farmacia 2011; 12-26.

[23] Salim AA, Chin YW, Kinghorn AD. Drug discovery from plants. 2008.

[http://dx.doi.org/10.1007/978-3-540-74603-4_1]

[24] Kayser O, Wim J, Eds. Biotechnology, MP From Basic Research to Industrial Applications. Quax Copyright© 2007 WILEY-VCH Verlag GmbH & Co. KGaA, Weinheim) 2007.

[25] Houghton PJ. Old yet new—pharmaceuticals from plants. J Chem Educ 2001; 78: 175-84. [http://dx.doi.org/10.1021/ed078p175]

[26] Máthé A. Genetic transformation of MAPs. Engineering StudiesMedicinal and Aromatic Plants of the World. Dordrecht: Springer Science+Business Media 2015.

[27] Balunas MJ, Kinghorn AD. Drug discovery from medicinal plants. Life Sci 2005; 78(5): 431-41. [http://dx.doi.org/10.1016/j.lfs.2005.09.012] [PMID: 16198377]

[28] Rates SK. Some traditional herbal medicines. Toxicon 2001; 39: 603-13. [http://dx.doi.org/10.1016/S0041-0101(00)00154-9] [PMID: 11072038]

[29] Brito AR. How to study the pharmacology of medicinal plants in underdeveloped countries. J Ethnopharmacol 1996; 54(2-3): 131-8. [http://dx.doi.org/10.1016/S0378-8741(96)01460-2] [PMID: 8953427]

[30] Souza Brito AR, Nunes DS. Ethnopharmacology and the sustainable development of new plant-derived drugs. Ciencia e Cultura. Journal of the Brazilian Association for the Advancement of Science 1997; 49(5/6): 402-8.

CHAPTER 2

Active Principles in MAPs

M. Paz Arraiza[1,*], **Carlos Calderón-Guerrero**[1], **José V. López**[1], **Silvia C. Guillén**[2], **Miguel A. Sarmiento**[3] and **Diego A. Meloni**[4]

[1] *Department of Forestry and Environmental Engineering and Management, Technical University of Madrid, Spain*

[2] *Ecology Department, Faculty of Agronomy, National University of Tucuman, Argentina*

[3] *Faculty of Forestry, National University of Santiago del Estero, Argentina*

[4] *Faculty of Agronomy and Agroindustries, National University of Santiago del Estero, Argentina*

Abstract: Pharmacological activity in plant-based drugs is centred on the presence of chemical compounds which we call active compounds. Active components are substances which are found in different parts and organs of plants, which change or modify the functions of human and animal organs and systems. There is a huge variety of active components, of which the most important are secondary metabolites such as essential oils, alkaloids, glycosides and phenolics. There are many classifications but we will use the following classification in four groups according to chemical structure: glycosides, phenolics, terpenoids, and alkaloids. These in turn are sub-divided into other groups according to their chemical structure. In this chapter, a classification, description function and examples of these active compounds is presented. We will make reference to secondary metabolites only, due to their significance as a source for active ingredients.

Keywords: Active compounds, Alkaloids, Anthraquinonic glycosides, Cardio-active glycosides, Coumarins, Cyanogenic glycosides, Diterpenes, Essential oils, Flavonoid glycosides, Flavonoids, Glycosides, Iridoids, Isoquinoleine, Isothio-cyanate glycosides, Lactones, Lignans, Phenolic acids, Phenolic glycosides, Phenolics, Polyphenols, Quinine, Quinones, Saponins, Tannins, Terpenoids, Tropane.

INTRODUCTION

Active components are substances which are found in different parts and organs of plants, and which may affect in several ways the functions of human and animal organs and systems. There is a huge variety of active components, of

* **Corresponding author María Paz Arraiza:** Department of Forestry and Environmental Engineering and Management, Technical University of Madrid, Madrid, Spain; Tel/Fax: +34 91 336 64 08; Email: paz.arraiza@upm.es

M. Paz Arraiza, A. González-Coloma, J. Burillo and C. Calderón-Guerrero (Eds.)

which the most important are secondary metabolites such as essential oils, alkaloids, glycosides and phenolics.

In this chapter we will make reference to secondary metabolites only, for these are the most important ones as a source for active ingredients.

Secondary metabolites do no result from primary metabolism and do not intervene in survival, growth and reproduction, but they are synthesised by plants as means for defence, adaptation, pollination, *etc.* There are many classifications (according to chemical structure, origin, action) but we will use the following classification in four groups according to chemical structure:

Glycosides: Alcohol glycosides, aldehyde glycosides, anthraquinone glycosides, cardioactive glycosides, coumarin and furanocoumarin glycosides, cyanogenic glycosides, flavonoid glycosides, isothiocyanate glycosides, lactone glycosides, phenolic glycosides, saponins.

Phenolic Compounds: Phenolic acids, coumarins, flavonoids, lignans, tannins.

Terpenoids: Monoterpenes, sesquiterpenes, diterpenes, essential oils, iridoids, lactones.

Alkaloids: Derived from tropane, quinine or isoquinoleine.

GLYCOSIDES

Glycosides are compounds which are made up of two parts: one is a sugar (*e.g.* glucose) and the other a non-sugar or aglucone, aglycone, or genine. The bond between each part is hydrolizable, and needs to be broken to activate the compound; this break is catalysed by enzymes contained in the plant itself. They are classified according to the structural features of the non-sugar or aglycone. The names end in –oside, although some prefer to use the traditional names ending –ine (*e.g.* digitoxin).

They make up the active components of many plants, and their pharmacological activity is basically due to the non-glucidic part. The most important groups are anthraquinone, cardiac, coumarin and cyanogenic glycosides and saponins [1].

Anthraquinone Glycosides

These consist of a molecule of sugar bonded to an anthracene derivative, more or less hydroxylated. They are used as laxatives and purgatives, and also are used as dyes. They are found in the Liliaceae, Rubiaceae, Leguminosae, Rhamnaceae and Polygonaceae families, among others. They are, for example, found in the following plants:

- Aloe juice, *Aloe vera, A. barbadensis* (Xanthorroaceae).
- Common buckthorn bark, *Rhamnus frangula* (Rhamnaceae).
- Purging buckthorn fruit, *Rhamnus cathartica* (Rhamnaceae).
- Rhubarb rhizomes, *Rheum rhabarbarum* (Polygonaceae).
- Sacred bark, *Rhamnus purshiana* (Rhamnaceae).
- Senna follicles and pods, *Cassia fistula* (Fabaceae).

Cardiac Glycosides

This group contains a number of active components which act directly on the cardiac muscle, and thus exert therapeutic influence on congestive cardiac insufficiency, or on cardiac arrhythmia. The group includes drugs such as digoxin, digitoxin, and ouabain. Nowadays, they are been studied in the prevention and/or treatment of cancer [2]. Some of the drugs which contain this type of substance are:

- Foxglove dry leaves, *Digitalis purpurea* and *D. lanata* (Scrophulariaceae).
- Dried scales of bulbs of sea squill, *Drimia maritima* (Asparagaceae).
- Oleander leaves, highly poisonous, *Nerium oleander* (Apocynaceae).
- Strophanthus seeds, *Strophantus gratus* and *Strophantus kombe* (Apocynaceae).

They share a common characteristic, which is that they act specifically on the heart; no synthetic substitute has been found for some of them. Cardiac glycoside drugs are highly toxic.

Coumarin Glycosides

These are aromatic molecules which decrease capillary permeability and increase capillary walls resistance, protecting from capillary weakness and acting as a tonic for the veins. Some have sedative properties, such as angelicyne. They may have hypnotic properties.

They are bitter compounds that activate hunger and digestion, and excite the tongue taste papilla. As a reflex reaction, they act on the stomach, increasing motility and helping to increase secretion. They should not be used when lactating as the active components pass into mother's milk. They have diuretic, anti-rheumatic, and anti-arthritic properties. They are used to treat the kidneys and the bladder diseases, or in cases of urine retention. They are found in Gentian (*Gentiana lutea,* Gentianaceae), which contains amarogentin, in the bark of Indian chestnut (*Aesculus hippocastanum,* Sapindaceae), and ash leaves (*Fraxinus excelsior,* Oleaceae).

Cyanogenic Glycosides

Some plants produce cyanhydric acid (in a process known as cyanogenesis), which causes lethal poisoning. They are found mainly in Rosaceae, Fabaceae, and Euphorbiaceae.

- Roots: tapioca, *Manihot utilissima* (Euphorbiaceae).
- Bark: cherry laurel, *Prunus laurocerasus* (Rosaceae).
- Fruit: elder, *Sambucus nigra* (Adoxaceae).
- Seeds: almonds, *Prunus amygdalus* (Rosaceae).

The sweet variety of almonds is edible, the bitter variety contains amygdaline. When crumbled and mixed with water, they give off a smell like benzaldehyde. If eaten they are toxic because they contain cyanide ions. 20-50 almonds are enough to kill an adult.

Saponins

Or saponosides, from the Latin *sapo* = soap. They are substances which produce foam in water solutions, and are naturally tensoactive. Many have haemolytic properties (disintegration of erythrocytes), and are highly toxic when injected into the blood. Their toxicity is lower when taken orally. They are toxic to cold-blooded animals specially fish [3].

There are triterpenoid and steroidal types. Triterpenoid saponins have anti-inflammatory, molluscicidal, anti-tussive, expectorant, analgesic and cytotoxic activities, such as ginsenosides, responsible for the pharmacological activity of ginseng (*Panax ginseng*) and the active triterpenoid saponins from liquorice (*Glycyrrhiza glabra*).

They are also found in horse chestnut (*Aesculus hippocastanum*), and in Asian centella (*Centella asiatica*). The steroidal saponins are important precursors for steroid drugs including anti-inflammatory agents, androgens, oestrogens and progestins [4]. They are found in Ruscus (*Ruscus aculeatus*), agave (*Agave sisalana*), and sapogenins and diosgenin in dioscorea (*Dioscorea spp.*), hecogenin from *Agave* and smilagenin from *Smilax*. Saponins from ginseng and liquorice exhibit many therapeutic effects [4].

PHENOLIC COMPOUNDS

Phenolic compounds are substances with a benzene nucleus supporting a hydroxyl group. They usually bond with sugars to form glycosides, but they may be found separate. There are simple and complex phenolics, and they are derived from two metabolic pathways, shikimic acid and malonic acid, the first one most important in vascular plants and the second one in bacteria and fungi [1]. Unbounded phenols provide important constituents of essences, such as thymol and its isomer, carvacrol (found in essential oil of thyme). Eugenol, for example, is antiseptic and also a local anaesthetic used in dentistry. Many phenols are found in an oxidized ether state in essential oils (estragole, myristicin, apiol, anethole). They range from very simple substances to very complex ones such as lignins and tannins. The main groups of phenolic compounds are phenolic acids, coumarins, flavonoids, lignans and tannins (Fig. **1**).

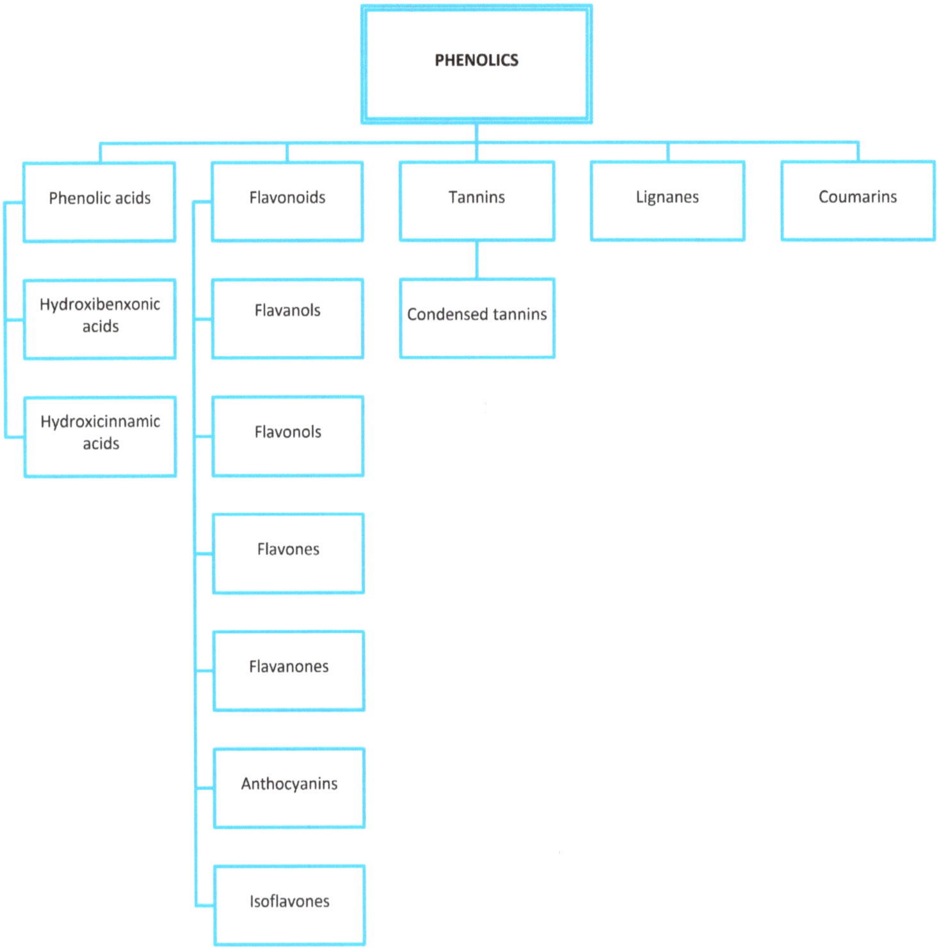

Fig. (1). Main groups of phenolic compounds.

Phenolic Acids

Phenolic acids are divided into two subgroups, the hydroxybenzoic and hydroxycinnamic acids. Hydroxybenzoic acids include gallic, p-hydroxybenzoic, protocatechuic, vanillic and syringic acids, with C6–C1 structure. Hydroxycinnamic acids are aromatic compounds with a C6–C3 structure, being caffeic, ferulic, p-coumaric and sinapic acids the most common. They have antioxidant activity, which increases with increasing degree of hydroxylation, and hydroxycinnamic acids present higher antioxidant activity than the corresponding hydroxybenzoic acids.

Flavonoids

Flavonoids are a major coloring component of flowering plants [5]. They are derived from phenyl-benzo γ pirone or phenyl chromone [6]. They are frequently found in the vegetable kingdom, usually in the form of glycosides. They are a very varied family of compounds, although all the final products are typically polyphenolic and soluble in water. There are several classes, including flavones, flavonols, flavanols, anthocyanines, chalcones, isoflavones and neoflavonoids, as well as two others, xhantones, and aurones. Of all these, the most interesting in terms of pharmacology are flavones, flavonols, and isoflavones, and their correspondent glycosides and anthocyanosides.

The major groups of flavonoids of nutritional interest are flavanols or catechins (from green tea), flavones (apigenin from celery), flavonols (quercetin, in red onion and apple), flavanones (naringenin from citrus), anthocyanidins (cyanidin from berries), and isoflavones (genistein and daidzein from soya beans) [7].

Among the medicinal plants whose activity is related to flavonoid content are passion flower (*Passiflora incarnata*), camomile (*Chamaemelum nobile*), yarrow (*Achillea millefolium*), liquorice (*Glycyrrhiza glabra*), ginkgo (*Ginkgo biloba*), milk thistle (*Sylibum marianum*) and white thorn (*Crataegus monogyna*).

The anthocyanoside group is responsible for the red and blue colouration of many flowers, fruits and leaves, thus helping pollination by attracting insects, such as cranberry, *Vaccinium mirtyllus* (Grossulariaceae), and blackcurrant, *Ribes nigrum* (Saxifragaceae).

Coumarins

Coumarins are benzo - α pyrones, a large group of phenolic active components found in medicinal plants, which share a chemical structure 2H-1-benzopyra--2-one known as coumarin. Substitutes of different chemical nature follow this structure which gives us different types of coumarins (simple coumarins, furanocoumarins, pyranocoumarins and the pyrone-substituted coumarins) [8]. They are found in some essential oils such as cinnamon bark oil, cassia leaf oil and lavender oil, and mainly in the Rutaceae and Umbelliferae families [8]. The most common one is umbelliferone. Practically all coumarins, with the exception of coumarin itself, have a hydroxilic substitute (OH) at C7, be it free, as in umbelliferone, or in combination (methyl, sugars, *etc.*) [9].

As a group, it's not particularly interesting in pharmacological terms. Nevertheless, we must mention its effects on the vascular system (both on arteries and veins) and its use as a treatment for certain ailments of the skin such as psoriasis, due to its photosensitizing properties. Some of them are:

- Coumarin, found in extracts of the melilot (*Melilotus officinalis*), is anti-inflammatory [10].
- Esculoside, found in Indian chestnuts (*Aesculus hippocastanum*) is both a tonic for the veins, used in haemorrhoid treatment [11]
- Visnadine is a dilator of the blood vessels found in Toothpick weed (*Ammi visnaga)* used as circulation improver [12]
- Dicumarol is an anticoagulant found in melilot (*Melilotus officinalis*) [13]
- Furanocoumarins are photosensitizing and are used to treat psoriasis. Sometimes they are used in sun creams as they enhance melanin production (photodynamic), for example, essential oil of bergamot (*Citrus bergamia*) [14].

Lignans

Lignans are molecules whose structure is the union of two units of phenyl propane (C6-C3). They are very common in the plant world. Podophyllotoxin, for example, is found in the podophylle rhizomes (*Podophyllum peltatum*) and is the fore-runner of two substances (etoposide and teniposide) used for anti-tumor therapy. It shows biological activity mainly as strong antiviral agent and as antineoplastic drug [15]. Silymarin is an antioxidant for liver and skin [12] obtained from the milk thistle (*Silybum marianum*).

Tannins

Tannins are complex substances which we cannot classify under one single chemical structure. They are no nitrogenised polyphenolic hydro-soluble substances, having a molecular weight between 500-3000. Apart from reacting in the classic phenol way, they precipitate gelatine, alkaloid salts, and heavy metals. They can be both hydrolysable and condensed [16].

Tannin is found mainly in the roots, bark, and sometimes in the leaves of plants. These compounds have antibacterial, astringent, and antiseptic properties. They are found in Ericaceae, Fabaceae, Rosaceae, and Salicaceae families.

Historically, they are substances used for tanning leather as they form hydrogen bridges with the fibres of collagen in the skin. As pharmaceutical items, tannins are used against diarrhoea and as an antidote in poisoning by heavy metals. Their use declined after the discovery of hepatotoxic effects of absorbed tannic acids. Recent studies have reported that tannins have anti-cancer and anti-HIV activities [4].

TERPENOIDS

Terpenoids are made up of a whole number of isoprene (C5) units. Two isoprene units make a monoterpene (C10). Depending on the number of carbons, they are classified as Monoterpenes (C10), Sesquiterpenes (C15), Diterpenes (C20), Triterpenes (C30), Tetraterpenes (C40), Polyterpenes (Cn). They include essential oils, iridoids, sesquiterpenic lactones, saponins, and cardiac glycosides.

Monoterpenes

Monoterpenes are the main constituents (90%) of essential oils. They are widely distributed in the plant kingdom and they are used in food, perfumes and human health care. Their pharmacological properties are antifungal, antibacterial, antioxidant, anticancer, antiarrhythmic, anti-aggregating, local anesthetic, antinociceptive, anti-inflammatory, antihistaminic and anti-spasmodic activities among many others [17, 18].

There are hydrocarbons such as limonene, the precursor to the main components of mint essences (*Mentha* spp., Lamiaceae) like carvone and menthol. α and β pinene are found in turpentine essence of *Pinus* spp. (Pinaceae) [19]. They have stimulant, decongestant, antiviral and antitumour properties [20].

Monoterpene alcohols like linalool are found in roses and lavender and are the main component of *Mentha arvensis*. Linalool gives tea, thyme, and cardamom leaves their taste. Menthol is responsible for the smell and taste of mint, which essence may contain up to 50% of this component. Geraniol, from scented geraniums (*Pelargonium* spp.), citronellol, from roses (*Rosa gallica*), borneol from rosemary, and santalol from sandalwood (*Santalum album,* Santalaceae) are also monoterpenic alcohols. They have antimicrobial, antiseptic, tonifying, spasmolytic properties [20].

Aldehyde monoterpenes, such as geranial, citral or citronellal are highly reactive compounds. They are found in abundance in citrus plants, and are responsible for their characteristic smell. They have spasmolytic, sedative, antiviral properties [20].

Phenolic monoterpenes are only found in a few species but are very powerful and irritating. The most important are thymol and carvacrol, which are found in thyme (*Thymus* spp.) and oregano (*Origanum* spp.), both of the Labiatae family [21]. Another important phenol is eugenol, which is found in many species, for example, clove essence. It is both a powerful bactericide and also anaesthetic, and is used in dentistry [22]. They have antimicrobial, irritant, immune stimulating properties [20].

Phenolic ethers are the main components of species such as celery and parsley (apiol), aniseed (anethol), basil and estragon (methylchavicol). Safrol is a component which is used extensively in the perfume industry and is found in the bark of the sassafras tree (*Sassafras albidum,* Lauraceae).

Ketones are produced by the oxidisation of alcohols and are fairly stable molecules. Carvone is found in *Mentha spicata.* Thuyone (*Thuja occidentalis*) and pulegone (*Mentha pulegium*) are fairly toxic and should never be used during pregnancy. Thuyone is found in absinthe (*Artemisia absinthium*), and in salvia (*Salvia officinalis*). They have mucolytic, cell regenerating and neurotoxic properties [20].

Ethers or monoterpene oxides are reactive and unstable. One example is bisabolol oxide found in camomile (*Matricaria chamomilla*). Another common ether is 1.8 –cineol (also known as eucalyptol), which is the main component of eucalyptus oil. It is an expectorant and mucolitic, and the main component of cough medicines. The aroma of eucalyptus oil varies depending on 1.8 –cineol content.

Monoterpene esters are formed from a reaction of a terpene alcohol with acetic acid. Their aroma is characteristic of the oils in which they are found. Lavender oil, for example, contains linalool in its ester, linalyl acetate. The relative abundance of both these components is a sign of good quality [20].

Sesquiterpenes

Sesquiterpenes, *e.g.* bisabolol, humulene and caryophyllene, are constituents of essential oils of many plants. Sesquiterpene lactones are known as bitter principles. They are found in families like the Asteraceae, Lauraceae and Magnoliaceae. They have anti-bacterial, anti-fungal, anthelmintic, anti-malarial and molluscicidal activity [4]. They are responsible for the bitter taste of many drugs such as those obtained from milk thistle (*Silybum marianum*), absinthe (*Artemisia absinthium*), or dandelions (*Taraxacum officinale*).

Diterpenes

Diterpenes have therapeutic applications. Taxol, from *Taxum brevifolia* and *T. baccata* (Taxaceae) and its derivatives are anti-cancer drugs. Forskolin, from the Indian Coleus plant (*Coleus forskohlii*), has anti-hypertensive activity. Stevioside from *Stevia rebaudiana* is a sweetening agent [4].

Triterpenes

Triterpenes are a group of active ingredients present in plants. While mono and sesquiterpenes are mostly found in plant essential oils triterpenes, are mainly found in balsams and resins. They are usually found as glycosides their bioactivity depending on the aglycone parts. The pharmaceutical applications of triterpenes are considerable, mainly anti-inflammatory, hepatoprotective, analgesic, antimicrobial, antimycotic, virostatic, immunomodulatory and tonic. Examples of these compounds are oleanane, ursane, lupine and dammarane [23].

Iridoids

These are monoterpenic compounds and the name comes from a type of Australian ant (*Iridomirmex*) whence iridodial, the simplest compound in this group, was first isolated. They are usually found in plants in the form of glycosides, in the Gentianaceae and Valerianaceae families. They are part of the active principle of some plants, such as the valepotriates of the valerian root (*Valeriana officinalis*), harpagosides of the harpagophite (*Harpagophytum procumbens*), oleuropeoside in olive leafs (*Olea europea*), gentiopicroside and amarogentin in the roots of the gentian (*Gentiana lutea*) [4].

ALKALOIDS

This is a group of natural products of particular interest to pharmacology. Within this group we can find substances which are toxic even in small doses. The first alkaloid to be isolated was morphine (Sertüner 1805) [24]. The name 'alkaloids' was hit upon in 1819, because of their basic nature. Although they were first isolated in the 19th century, their complexity meant that their structure was not established until later.

This was the case with strychnine (vomic nut seed). It was isolated in 1819. In 1870 its structure was established approximately. In 1889 it was produced synthetically. In 1946 its structure was finally determined [24].

Our knowledge of natural alkaloids has progressed thanks to the development of new techniques of separation and identification. In 1930, more than 300 were isolated and the structure of 200 was established. In 1950 more than 1,000 were isolated, in 1973, between 5 and 6,000. Nowadays more than 27,000 alkaloid structures are known [25].

There are several types, depending on the molecule from which they derive (tropane, quinoleine or isoquinoleine).

Drugs with Alkaloids Derived from Tropane

These are parasympathomimetic substances or antagonists of acetylcholine (synaptic transmitters). They have the following effects on the central nervous system: in large doses, they cause vertigo, delirium, hallucinations. In very large doses, it causes death by respiratory paralysis. The drugs potency is higher in hyosciamine, followed by atropine and scopolamine [26].

Cocaine (leaves of the *Erythroxylum coca*, Erythroxilaceae) is a powerful local anaesthetic (it lengthens the effects of anaesthesia by constricting blood vessels). It excites the central nervous system. In extreme doses, it causes death by respiratory paralysis. Lysergic acid is produced by a fungus, *Claviceps purpurea*, which grows around the ovary of rye. Its alkaloids have been used for treating migraine headaches and Parkinson's disease among others [27]. Caffeine is the alkaloid found in coffee grains (*Coffea arabica*), tea leaves (*Camellia sinensis*) or mate (*Ilex paraguariensis*). Theobromine is a related molecule found in cocoa (*Theobroma cacao*). Atropine is found in leaves and flowers of deadly nightshade, *Atropa belladonna* (Solanaceae). Daturine comes from Jimson weed leaves *Datura stramonium* (Solanaceae). Hyosciamin is found in henbane leaves. *Hyoscyamus niger* (Solanaceae), it is used to treat Parkinson’s disease [26].

Drugs with Alkaloids Derived from Quinoleine

The drug quinine is the dried bark of the trunk, branches and roots of the genus *Cinchona* spp. (Rubiaceae). The most important species are *Cinchona pubescens* (red quine), *C. ledgeriana* (ledgerine quine), *C. calisaya* (yellow quine), *C. hybrida* and *C. robusta*. Their main effect is anti-malarial due to the quinine, which is toxic to the protozoa and paramecia especially for the plasmodium which produces malaria (*Plasmodium vivax*, *P. malaria*) [28] [29]. Quinoline alkaloid is a cardiac depressant [26].

Drugs with Alkaloids Derived from Isoquinoleine

Opium is the latex obtained from incising unripe pods of the different varieties of

Papaver somniferum (Papaveraceae). Morphine and codeine are obtained from it [30]. Even in low doses it is analgesic, and decreases the perception of pain; particularly useful against persistent, acute pains. It is sedative to the central nervous system, leading to an unpleasant drowsiness, produces a sense of wellbeing and euphoria. Repeated administration causes addiction, tolerance, and physical and psychological dependence. It causes respiratory depression and bradicardia. In the intestines it is an emetic and diminishes peristalsis. Overdoses cause death by cardiac and respiratory arrest [26]. Colchicine from *Colchicum autumnale* is an antigout drug. Emetine, from *Cephaelis ipecacuanha* is antiamoebic. Galanthamine from *Galanthus nivalis* and *Leucojum aestivum* is a cholinesterase inhibitor used to treat Parkinson's disease. Tubocurarine from *Chondodendron tomentosum* is a skeletal muscle relaxant [26].

CONFLICT OF INTEREST

The authors declare no conflict of interest, financial or otherwise.

ACKNOWLEDGEMENTS

Declared none.

REFERENCES

[1] Ávalos García A, Pérez-Urria Carril E. Metabolismo secundario de plantas. Reduca (Biología) 2011; 2(3)

[2] Cragg GM, Newman DJ. Plants as a source of anti-cancer agents. J Ethnopharmacol 2005; 100(1-2): 72-9.
[http://dx.doi.org/10.1016/j.jep.2005.05.011] [PMID: 16009521]

[3] Sparg SG, Light ME, van Staden J. Biological activities and distribution of plant saponins. J Ethnopharmacol 2004; 94(2-3): 219-43.
[http://dx.doi.org/10.1016/j.jep.2004.05.016] [PMID: 15325725]

[4] Gurib-Fakim A. Medicinal plants: traditions of yesterday and drugs of tomorrow. Mol Aspects Med 2006; 27(1): 1-93.
[http://dx.doi.org/10.1016/j.mam.2005.07.008] [PMID: 16105678]

[5] Yao LH, Jiang YM, Shi J, *et al.* Flavonoids in food and their health benefits. Plant Foods Hum Nutr 2004; 59(3): 113-22.
[http://dx.doi.org/10.1007/s11130-004-0049-7] [PMID: 15678717]

[6] Hässig A, Liang WX, Schwabl H, Stampfli K. Flavonoids and tannins: plant-based antioxidants with vitamin character. Med Hypotheses 1999; 52(5): 479-81.
[http://dx.doi.org/10.1054/mehy.1997.0686] [PMID: 10416956]

[7] Birt DF, Jeffery E. Flavonoids. Adv Nutr 2013; 4(5): 576-7.
[http://dx.doi.org/10.3945/an.113.004465] [PMID: 24038263]

[8] Jain P, Joshi H. Coumarin: Chemical and pharmacological profile 2012.

[9] Murray RD, Méndez J, Brown SA. The natura coumarins: Occurrence, Chemistry and Biochemistry. John Wiley & Sons 1982.

[10] Pleşca-Manea L, Pârvu AE, Pârvu M, Taămaş M, Buia R, Puia M. Effects of *Melilotus officinalis* on acute inflammation. Phytother Res 2002; 16(4): 316-9. [http://dx.doi.org/10.1002/ptr.875] [PMID: 12112285]

[11] Beylot G. La maladie hémorroïdaire. Actual Pharm 2008; 47(472): 43-6. [http://dx.doi.org/10.1016/S0515-3700(08)70262-3]

[12] Bhokare SG, Dongaonkar CC, Lahane SV, Salunke PB, Sawale VS, Thombare MS. Herbal novel drug delivery-. RE:view 2016.

[13] Link KP. The anticoagulant from spoiled sweet clover hay. Harvey Lect 1944; 39: 162-216.

[14] Ebermann R, Alth G, Kreitner M, Kubin A. Natural products derived from plants as potential drugs for the photodynamic destruction of tumor cells. J Photochem Photobiol B 1996; 36(2): 95-7. [http://dx.doi.org/10.1016/S1011-1344(96)07353-8] [PMID: 9002245]

[15] Canel C, Moraes RM, Dayan FE, Ferreira D. Podophyllotoxin. Phytochemistry 2000; 54(2): 115-20. [http://dx.doi.org/10.1016/S0031-9422(00)00094-7] [PMID: 10872202]

[16] Serrano J, Puupponen-Pimiä R, Dauer A, Aura AM, Saura-Calixto F. Tannins: current knowledge of food sources, intake, bioavailability and biological effects. Mol Nutr Food Res 2009; 53(2) (Suppl. 2): S310-29. [http://dx.doi.org/10.1002/mnfr.200900039] [PMID: 19437486]

[17] Kozioł A, Stryjewska A, Librowski T, *et al.* An overview of the pharmacological properties and potential applications of natural monoterpenes. Mini Rev Med Chem 2014; 14(14): 1156-68. [http://dx.doi.org/10.2174/1389557514666141127145820] [PMID: 25429661]

[18] Guimarães AG, Quintans JS, Quintans LJ Jr. Monoterpenes with analgesic activity--a systematic review. Phytother Res 2013; 27(1): 1-15. [http://dx.doi.org/10.1002/ptr.4686] [PMID: 23296806]

[19] Wrolstad RE, Jennings WG. Volatile constituents of black pepper. III. The monoterpene hydrocarbon fraction. J Food Sci 1965; 30(2): 274-9. [http://dx.doi.org/10.1111/j.1365-2621.1965.tb00301.x]

[20] Pengelly A. The constituents of medicinal plants: an introduction to the chemistry and therapeutics of herbal medicine| NOVA. The University of Newcastle's Digital Repository 2004.

[21] Stahl-Biskup E. The chemical composition of *Thymus* oils: a review of the literature 1960–1989. J Essent Oil Res 1991; 3(2): 61-82. [http://dx.doi.org/10.1080/10412905.1991.9697915]

[22] Valnet J, Tisserand RB. The practice of aromatherapy. Saffron Walden: CW Daniel Co. Ltd. 1980.

[23] Dzubak P, Hajduch M, Vydra D, *et al.* Pharmacological activities of natural triterpenoids and their therapeutic implications. Nat Prod Rep 2006; 23(3): 394-411. [http://dx.doi.org/10.1039/b515312n] [PMID: 16741586]

[24] Schiff PL Jr. Opium and its alkaloids. Am J Pharm Educ 2002; 66(2): 186.

[25] Wink M. Modes of action of herbal medicines and plant secondary metabolites. Medicines 2015; 3: 251-86. [http://dx.doi.org/10.3390/medicines2030251]

[26] Raskin I, Ribnicky DM, Komarnytsky S, *et al.* Plants and human health in the twenty-first century. Trends Biotechnol 2002; 20(12): 522-31. [http://dx.doi.org/10.1016/S0167-7799(02)02080-2] [PMID: 12443874]

[27] Schiff PL Jr. Ergot and its alkaloids. Am J Pharm Educ 2006; 70(5): 98. [http://dx.doi.org/10.5688/aj700598] [PMID: 17149427]

[28] Schmeller T, Wink M. Utilization of alkaloids in modern medicine. In Alkaloids Springer, US 1998; pp. 435-59.

[29] Werner L. Quinine's: Feverish tales and trails. Americas 2003; 55(5): 24.

[30] Blakemore PR, White JD. Morphine, the Proteus of organic molecules. Chem Commun (Camb) 2002; 11(11): 1159-68.
[http://dx.doi.org/10.1039/b111551k] [PMID: 12109065]

CHAPTER 3

Industrial Uses of MAPs: Cosmetic Industry

M. Paz Arraiza[1,*], **Carlos Calderón-Guerrero**[1], **Silvia C. Guillén**[2] and **Miguel A. Sarmiento**[3]

[1] *Department of Forestry and Environmental Engineering and Management, Technical University of Madrid, Spain*

[2] *Ecology Department, Agronomy Faculty, National University of Tucuman, Argentina*

[3] *Faculty of Forestry, National University of Santiago del Estero, Argentina*

Abstract: Plants generate a considerable amount of antioxidants, preservatives, and synthetic colours. There is much demand for these in the cosmetics industry. The perfume sector also uses important quantities of aromatic plants. The dermatological properties of plants are many and varied: tonics, astringents, anti-inflammatory, antiseptics, anti-scarring, cleansers, moisturisers, relaxants, emulsifiers, decongestant, refreshing, *etc.* The use of different plants is determined by their physiological activity, which varies from plant to plant, so it is possible to find a plant for nearly all our aesthetic needs. There is also a wide range of formats of cosmetic preparations. We can find preparations of plant extracts in creams, emulsions, lotions, gels, oils, soaps, deodorants, *etc.* Regulations regarding these products apply to raw materials for use in cosmetics. They also refer to the vegetable extracts used in the manufacture of cosmetic products.

Keywords: Cleansing, Conditioning, Cosmetics, Creams, Dental care, Essential oils, Extracts, Hair care, Hair tonic, Healing, Lotions, Moisturizers, Moisturizing, Nourishing, Natural cosmetics, Natural ingredients, Perfumes, Protection, Shaving cream, Skin care, Soap.

INTRODUCTION

Many plants are used in the cosmetic industry, and by cosmetic industry we understand skin care products, hair care products, dental care, additives and preservatives and perfumes.

We can also make three categories of products obtained from plants which are used in the cosmetic industry, which is to say, vegetable oils, essential oils and botanical extracts.

* **Corresponding author María Paz Arraiza:** Technical University of Madrid. Department of Forestry and Environmental Engineering and Management, Madrid, Spain; Tel/Fax: +34 91 336 64 08; Email: paz.arraiza@upm.es

M. Paz Arraiza, A. González-Coloma, J. Burillo and C. Calderón-Guerrero (Eds.)

There are many plants which produce vegetable oils used in cosmetics, such as argan oil (*Argania spinosa*), avocado seed oil (*Persea americana*), castor oil (*Ricinus communis*), cocoa butter (*Cocos nucifera*), jojoba oil (*Simmondsia chinensis*), olive oil (*Olea europaea*), rose hip oil (*Rosa rubiginosa*), sesame seed oil (*Sesamum indicum*), shea butter oil (*Butyrospermum parkii*), soya oil (*Glycine max*), sunflower seed oil (*Helianthus annuus*), sweet almond oil (*Prunus dulcis*) and wheat germ oil (*Triticum vulgare*).

We can highlight many plants which produce extracts used in the cosmetic industry, such as aloe (*Aloe vera*), butcher´s broom (*Ruscus aculeatus*), calendula (*Calendula officinalis*), chamomile (*Chamomilla recutita*), ginkgo (*Gingko biloba)*, ginseng (*Zingiber officinale*), green tea (*Camellia sinesis*), Indian chestnut (*Aesculus hippocastanum*), ivy (*Hedera helix*), oats (*Avena sativa*) and rosemary (*Rosmarinus officinalis*) among many others.

Some examples of plants which produce essential oils used in cosmetics are frankincense (*Boswellia serrata*), helychrisum (*Helychrisum stoechas*), lavender (*Lavandula* spp.), mints (*Mentha* spp.), moringa oil (*Moringa oleifera*), orange blossom (*Citrus aurantium*), rose (*Rosa damascena*), rose geranium (*Pelargonium graveolens*), sage (*Salvia officinalis*), sandalwood (*Santalum album*), tea tree (*Melaleuca alternifolia*), thyme (*Thymus* spp.), wintergreen (*Gaulteria procumbens*) and ylang ylang (*Cananga odorata*) essential oils.

These plants are used in several applications in the cosmetic industry that we could summarize in the following four main groups.

Skin Care: They are perfect for active ingredients of any product for skin care. Bath products, beauty treatments, body deodorants, cosmeceuticals, depilatory products, make-up, products for manicure and pedicure, products for shaving and aftershaves, skin treatment, sun-block and tanning preparations.
Hair Care: Hair care products, lotions, oils, anti-dandruff treatments.
Dental Care: Teeth and mouth hygiene products. Toothpaste, mouthwash.
Preservatives: Many essential oils have antimicrobial activity, and are thus added to synthetic cosmetics as support agents.
Fragrances: Perfumery is the main industrial cosmetic use of essential oils, even if synthetic products are more stable and last longer. Essential oils, perfumes and colognes.

MAPS USED FOR SKIN CARE

Cosmetics improve skin tone, texture, radiance, and reduce wrinkling [1]. Cosmetic products can be solid, such as powders, masks, *etc.*, semisolid, such as creams, ointments, liniments, wax creams, pastes, *etc.*, and liquids, such as lotions, cleansing milk, sprays, *etc.*

According to Ribeiro *et al.* [2] the most important effects of a cosmetic plant product are antioxidant activity (oxidation is the cause of skin ageing, due to UV radiation from sunlight), tyrosinase inhibition effect (over activity of tyrosinase leads to over-production of melanin, which may induce melanoma) and antimicrobial effect (the cosmetic industry is interested in replacing synthetic antimicrobials in topical products due to resistance to synthetic antimicrobials).

Basic skin care processes are:

- Cleansing and conditioning (lotions, de-makeup, exfoliating, tonics, masks).
- Hydration and nutrition (creams and lotions).
- Protection (creams, balms, sun filters).

Beneficial aspects of medicinal plants on skin include healing of wounds and burn injuries; antifungal, antiviral and antibacterial activity against skin infections such as acne, herpes and scabies; activity against inflammatory/immune disorders affecting skin (psoriasis); and promoting activity against skin cancer [3]. But plants also present side effects by irritant chemicals in plant sap (Ranunculaceae, Euphorbiaceae and Compositae families); or photodermatitis caused by furocoumarins, and subsequent exposure to UV light (Umbelliferae and Rutaceae plant families); and hypersensitivity contact reactions in individuals who are sensitive or allergic to plant products, such as peanuts [3].

Cleansing and Conditioning

Some plants reported to have cleansing action are Burdock (*Arctium lappa*), which is anti-inflammatory, antibacterial and antioxidant, and a helpful cleanser for eczema and acne; Lavender essential oil is calming, soothing and antiseptic. Calendula (*Calendula officinalis*) hydrates dry skin, helps heal wounds and relieves the side effects of radiotherapy. Feverfew (*Tanacetum parthenium*) reduces redness and inflammation from acne, rosacea or sunburn.

Distilled waters such as rose water, witch hazel (*Hammamelis virginiana)* water and orange blossom water are used for tonics due to their astringent, tonic, anti-irritant and anti-couperosis actions [4], also in face tonic lotions, after-shave lotions, products for sensitive skin and products for mucosa and inflamed skin [5]. Witch hazel removes excess of oil and contains tannins that relieve burns, bug bites and skin irritation. Rose (*Rosa damascena)* has calming, soothing, toning and astringent properties [1]. Rosemary (*Rosmarinus officinalis)* contains rosmarinic acid which is antioxidant and anti-inflammatory [5]. It is a sebum regulator for oily skin. It calms irritations, pimples, acne and roughness.

Cucumber (*Cucumis sativus)* water extract of fruits and seeds protects skin from sunburn [1].

Other cleansing plants are lemon (*Citrus limon),* nettle *(Urtica dioica)* as effective astringents [1] and soapwort (*Saponaria officinalis)* as a gentle effective cleanser thanks to its saponins [6].

Protection and Healing

Leaf-juice of aloe (*Aloe vera*) is applied on skin for smoothness, healing, controlling skin burn, sun burn and injury. It is used in moisturizers, lotions, creams, hair tonic, shaving creams, etc [1]. According to Shweta G. Gediya [7], other plants for skin protection are green tea (*Camellia sinensis*), calendula (*Calendula officinalis*) and turmeric (*Curcuma longa*). Green tea has a potent free radical with scavenger activity when applied to the skin [7], antioxidant and anti-inflammatory [4] before exposure to sunlight [8]. Ginseng, *Zingiber officinalis*, inhibits the oxidation of L tyrosine [9]. It is used in nutritive creams, anti-wrinkle and anti-aging cosmetics [5]. Tea tree (*Melaleuca alternifolia*) has antifungal, antiviral, and antibacterial properties against acne, herpes and scabies [5].

Lavender (*Lavandula* spp.) has calming, soothing and antiseptic properties. Feverfew (*Tanacetum parthenium*) reduces redness and inflammation from acne, rosacea and sunburn. Aloe (*Aloe vera*) has emollient, hydrating, moisturizing, epytellizing and anti-inflammatory properties. It speeds up the healing of wounds and injuries, skin and sun burns, so it is useful for sun protection and after sun [1, 5, 7, 10]. Calendula (*Calendula officinalis*) and chamomile (*Matricaria camomilla*) are skin-hydrating, heal wounds and relieve side effects of radiotherapy treatments [7]. Arnica (*Arnica montana*) is antioxidant and has a cytoprotective effect in treatment of skin disorders [11].

Moisturizing and Nourishment Plants

Lemon (*Citrus limon*) is a potential source of vitamin C; oil is used in various preparations to reduce skin itching and to nourish skin; the pulp left after extraction of juice is useful for facial ingredients [14].

According to Shweta G. Gediya [7], Golden Root (*Rhodiola rosea*) is an anti-aging plant due to its natural antioxidants. Other anti-aging plants are ginkgo (*Ginkgo biloba*), which strengthens small capillary veins), and carrot (*Daucus carota*) seed oil, due to its vitamin A content.

Rose geranium (*Pelargonium graveolens*), sandalwood (*Santalum album*) and YlangYlang (*Cananga odorata*) are used in beauty creams for skin beautification and protection from sunburn; anti-oxidant. Juniper (*Juniperus communis*) is used for skin rejuvenation. Rose (*Rosa damascena*) is used in skin creams, lotions and ointment for beautification, smoothness and protection from sunburns. Fatima Turmeric (*Curcuma longa*) rhizome powder possesses anti-inflammatory and anti-oxidant properties; used in face creams and ointments. Table **1** shows a compilation of some of the plants used for skin care.

Table 1. Some plants used for skin care.

Action	Plant name		Part used
Cleansing	Burdock	*Arctium appa*	Powdered root
	Lemon	*Citrus limon*	Essential oil
	Nettle	*Urtica dioica*	Leaves extract
	Soapwort	*Saponaria officinalis*	Flowers extract or water
	Witch hazel	*Hammamelis virginiana*	Flowers extract or water
Conditioning	Cucumber	*Cucumis sativa*	Fruit juice or extract
	Rose	*Rosa damascena*	Flowers extract or water
	Rosemary	*Rosmarinus officinalis*	Essential oil; leave extract
Protection	Green tea	*Camellia sinensis*	Leaves extract
	Ginkgo	*Ginkgo biloba*	Leaves extract
	Ginseng	*Zingiber officinalis*	Powdered root
	Tea tree	*Melaleuca alternifolia*	Essential oil
Healing	Aloe	*Aloe vera*	Leaves gel
	Arnica	*Arnica montana*	Flowers extract or water
	Calendula	*Calendula officinalis*	Flowers extract or water
	Chamomile	*Matricaria recutita*	Flowers extract or water
	Feverfew	*Tanacetum parthenium*	Flowers extract or water
	Lavender	*Lavandula spp*	Essential oil
Nourishing	Juniper	*Juniperus communis*	Essential oil; leave extract
	Rose	*Rosa damascena*	Flowers extract or water
	Rose Geranium	*Pelargonium*	Flowers extract or water
	Sandalwood	*Santalum album*	Wood essential oil
	YlangYlang	*Cananga odorata*	Flowers essential oil
Moisturizing	Comfrey	*Symphytum officinale*	Leaves and root extract
	Marshmallow	*Althaea officinalis*	Flowers extract or water
	Nettles	*Urtica dioica*	Leaves extract
	Turmeric	*Curcuma longa*	Powdered root

Vegetable Carrier Oils

The former plants produce mainly essential oils or herbal extracts used in skin products, but they are almost always accompanied by carrier oils that on the one hand provide the basis for the cosmetic properties and on the other, provide properties as well, such as moisturizing, protecting or nourishing. We cite here some of those most used in the cosmetic industry.

- Almond oil (*Prunus dulcis*). Moisturizer, emollient, soothing. Very widely used due to its high yield and versatility.
- Apricot kernel oil (*Prunus armeniaca*). Moisturizer, emollient.
- Argan oil (*Argania spinosa*). Skin moisturising, nourishment and regeneration.
- Avocado seed oil (*Persea americana*). Regenerating and restructuring dry skin.
- Cocoa butter *(Theobroma cacao).* Cellular regenerating, emollient
- Castor oil (*Ricinus communis*). Antioxidant, soothing, emollient.
- Coconut butter (*Cocos nucifera*). Hair and skin moisturising and nourishment.
- Jojoba wax (*Simmondsia chinensis*). Protecting, soothing, emollient, sebum balancing in facial anti-aging treatments. Non-greasy lubricant agent [12].
- Linseed oil (*Linum usitatissimum*). Moisturizer, emollient.
- Olive oil (*Olea europaea*). Antioxidants, moisturizing, cellular regeneration.
- Rapeseed oil (*Brassica napus*). Moisturizer, emollient.
- Rose hip oil (*Rosa rubiginosa*). Heals wounds. Anti-aging and whitening [4, 13].
- Sesame seed oil (*Sesamum indicum*). Skin protection and rejuvenation.
- Shea butter oil (*Butyrospermum parkii*). Moisturizing, anti-inflammatory.
- Sunflower seed oil (*Helianthus annuus*). Moisturizer, emollient.
- Wheat germ oil (*Triticum vulgare*). Nourishing for aging skin. It stimulates collagen synthesis and strengthens hair.

MAPS USED FOR HAIR CARE

Hair care comprises different uses such as tonics and hair care preparations, dyes, shampoos and anti-dandruff products among others. Many plants such as rosemary and nettles have been used for hair care since long ago.

Tonics and Hair Care Preparations

According to several authors [1, 14, 15] the following plant extracts have been used for hair care products.

Amla *(Phyllanthus emblica,* Euphorbiaceae*)* fruit extract is used in oils for promotion of hair growth [16]. Centella *(Centella asiatica,* Apiaceae) whole plant extract is used for the growth and maintenance of hair [17]. Cluster fig *(Ficus* *racemosa,* Moraceae) aerial root powder is mixed with coconut oil for massage to check falling hair [18]. Rosemary (*Rosmarinus officinalis*) glycolic extract is recommended in concentrations higher than 5% for hair lotions with stimulating and purifying properties [19]. Safflower (*Carthamus tinctorius,* Asteraceae) alcoholic extract is used in hair tonics. Thyme *(Thymus serpyllum,* Lamiaceae) whole herb extract is useful for preparing hair tonics [20].

Hair Dyes

Bastard myrobalan (*Terminalia bellirica,* Combretaceae) seed extract and oil is good for hair dyeing preparation. As well as the black myrobalan (*Terminalia chebula*), it is used as an anti-dandruff hair product [7]. Henna (*Lawsonia inermis,* Lythraceae) leaf paste is used for hair dyeing and nourishment [21]. Kuth (*Saussurea lappa,* Asteraceae) roots extract is used in hair dyeing. Sage (*Salvia officinalis,* Lamiaceae) aqueous extract is used as hair conditioner [22]. Walnut (*Juglans regia*, Juglandaceae) leaf and husk of fruits is used for hair dyeing [23].

Shampoo

Soapnut *(Sapindus mukorossi,* Sapindaceae). Due to its content in saponins, sugars and mucilage, the fruit coat extract is used in herbal shampoo as a hair cleanser.

Hair Oils

Coconut *(Cocos nucifera,* Arecaceae) kernel oil is well-established hair oil, and is used as such or as a basic raw material for preparing hair oils and tonics [15]. Sesame *(Sesamum indicum,* Pedaliaceae*)* seed oil is one of the major sources of hair oils, and is used as such or as a base for preparing specific hair oils.

Anti-dandruff Plants

According to Shweta G. Gediya *et al.* [7], some plants used to combat dandruff are neem (*Azadirachta indica*), camphor (*Cinnamomum camphora*), henna (*Lawsonia inermis*), Indian gooseberry (*Phyllanthus emblica*), rosary pea (*Abrus precatorius)* and sweet flag (*Acorus calamus*). Rosemary (*Rosmarinus officinalis*) and nettle (*Urtica dioica*) are also used as anti-dandruff agents.

MAPS USED FOR DENTAL CARE

Some authors [24] cite that the extracts of neem (*Azadirachta indica*), holy basil (*Ocimum sanctum*), curry tree (*Murraya koenigii*), acacia (*Acacia nilotica*), river red gum (*Eucalyptus camaldulensis*), roselle (*Hibiscus sabdariffa*), mango (*Mangifera indica*), guayaba (*Psidium guajava*), cyme rose (*Rosa indica*), and aloe (*Aloe barbadensis*) have been found to inhibit certain dental caries and periodontal pathogens.

Other authors [25] report the following plants as emerging trends of herbal care in dentistry: myrrh (*Commiphora myrrha*) promotes healing in cases of pyorrhea; peppermint (*Mentha piperita*) relieves toothache; red clover (*Trifolium pratense*) mouthwash heals irritated and diseased gums; rosemary (*Rosmarinus officinalis*) mouthwash for gum disease and bad breath; sanicle (*Sanicula europaea*) is a powerful antioxidant; shepherd's purse (*Capsella bursa-pastoris*) stops bleeding after tooth extraction; thyme (*Thymus vulgaris*) for oral herpes. It contains fluorine used in toothpaste; tea tree oil (*Melaleuca alternifolia*) relieves gum inflammation and dissolves the necrotic pulp tissue in root canal treatment; violet (*Clematis virginiana)* helps relieve the pain and tenderness from sores caused by oral cancer; wintergreen (*Gaultheria procumbens*) mouthwash is an excellent astringent and antiseptic; yarrow (*Achillea millefolium*) is used to treat hemorrhages, ulcers and to improve blood clotting, promoting healing of cuts in the mouth due to surgery, teeth cleaning and braces.

MAPS USED AS NATURAL PRESERVATIVES IN COSMETICS

Nowadays, the cosmetic industry, looking towards more natural and chemical-free products, is searching for raw materials of plant origin that can replace the synthetic ones traditionally used.

Preservation properties of antioxidants of natural origin are in general lower than those of synthetic compounds, but they have the advantage of not inducing secondary toxic effects. Flavonoids are a family of compounds with antioxidant action that prevent rancidity of oils and fats in creams. Plant extracts containing these compounds and thus used in cosmetics are grapefruit, rosemary and essential oils of tea and thyme [26].

Some extracts (*Matricaria chamomilla, Aloe vera, Calendula officinalis*) and essential oils (*Lavandula officinalis, Melaleuca alternifolia, Cinnamomum zeylanicum*) have been studied as compared with methylparaben [27]. These studies show that tested extracts and essential oils have the same preservative action in cosmetics as those found in methylparaben, providing guarantee of microbiological purity of the cosmetic in use and storage.

MAPS USED AS FRAGRANCES (PERFUMERY)

History of Perfumery

To find the origins of perfume we must travel back to Ancient Egypt. Incense (the word is of Latin origin and means “through smoke”) and myrrh were the basic ingredients used by the Egyptians to aromatise ceremonies and healing rituals, but also as a beauty complement for women. Furthermore, they had a complete process for the preparation of perfumes which consisted in impregnating rose leaves with oil to preserve their fragrance. They would also macerate aromatic plants in oil and then drain the oil onto a cloth.

Later, the Greeks inherited these techniques and proceeded to improve them. The Romans imported raw materials such as myrrh and incense from Arabia and brought other substances from India.

The fall of the Roman Empire brought about a decline in the use of perfumes in the West. The custom lived on only in Arabia, where plant distillation techniques were developed. Baghdad became the city of fragrances. Other ingredients were discovered, such as musk, which they mixed into the mortar used to build palaces and mosques.

During the middle ages, the Crusades saw a rise in commerce and trade between East and West and perfume was rediscovered. However, its general use did not come until the XVI century when Catalina de Medici imposed the perfume fashion in Paris.

Later on, the Industrial Revolution saw large-scale commercialisation of perfume. Perfumers specialised in chemistry so as to develop better products, and perfumes stopped being a fashion to become a basic requisite for the beauty criteria of the times.

At the beginning of the XX century, perfumes were given away in clothes shops, but gradually they began to be sold until the first exclusive perfume companies appeared.

- In France, Roger & Gallet (1862), Chanel (1910) and Christian Dior (1947).
- In the United Kingdom, K. Yardley (1770).
- In the United States, Elizabeth Arden (1910).
- In Spain, Gal (1899), Puig (1914) and Myrurgia (1919).

With time, better techniques developed, and new ingredients for the making of perfumes were discovered. Nowadays, it is an important industry with millions of dollars in turnover and which designers depend on more even than on clothes. Table **2** shows a compilation of perfumes through history.

Table 2. History of perfumes [28].

Date	Perfume	Origin	Composition
1800 BC	**Kifi**	Egypt	A paste made of wine, grapes, herbs, resins, incense, myrrh, honey and juniper.
400 BC	**Kypros**	Greece	Sweet wine, cardamom, aspalathos.
50 BC	**Ciprés**	Rome	Storax, labdanum and hemp.
400 AC	**Khaluq**	Arabia	Women ointment with orange blossom, rose, cubeb, cardamom and sesame.
600 AC	**Galilla**	Arabia	Luxury perfume of musk, ambergris, camphor and moringa.
800 AC	**Nadd**	Arabia	For burning. Aloe, ambergris, incense and gum Benjamin.
1370 AC	**Queen of Hungary**	Europe	Rosemary, marjoram, mint in alcohol. Later on lemon, Lily and lavender added.
1380 AC	**Eau del Carmen**	Spain	Lemon balm, lemon peel, orange blossom, spices and alcohol.
1714 AC	**Eau de Cologne**	France	Bergamot, mint, lemon, orange, petit grain, neroli, rosemary, rose and musk.
1910 AC	**English Lavender**	U. K.	Lavender, rosemary and herbs.

The Structure of Perfumes

For a long time, perfumes were thought to be evolving. But their fragrance and intensity were only due to evaporation, by this we mean volatility, and therefore to their gradual wearing off.

The 20th century saw the start of perfume preparation using several fragrances. The first to introduce this system of perfume preparation was André Guerlain in 1899. From then onwards, a system known as the "pyramid" or "three layer system" was generally used. This structure divides a perfume's fragrances into three types: high, medium, and low.

High Notes: These correspond to fragrances which evaporate faster because they come from more volatile substances. They last a very short time and are used to make the first impression of the fragrance. Among the high notes we find citrus, such as lemon, orange, tangerine, and lavender.

Middle Notes: These emerge when the high note has disappeared and are basically the "heart" of the perfume. They last about four hours and contain the main fragrances of a perfume. They include fragrances such as cinnamon, moss, herbal and floral essences

Finally, the **Low Notes:** These correspond to the least volatile fragrances. Their purpose is to "fix" the perfume and give it a global harmony. They are the longest lasting and may last up to two days. They include fragrances such as spices, tobacco, oak moss, jasmine, amber, and musk.

How long a perfume lasts depends on the environmental temperature and on body chemistry as well as on fragrance concentration. The higher the concentration of essential or fragrant oils, the more expensive the perfume is.

Types of Perfume according to their Concentration in Essential Oils

- **Perfume:** This is the highest concentration of a fragrance. It usually contains 20% of active ingredients, essential oils or fragrances. It lasts from 4 to 7 hours.
- **Eau de perfume:** It contains about 10% of essential oils or fragrances and once applied to the body, lasts between 3 and 5 hours.
- **Eau de toilette:** It contains about 5% concentration and lasts less than 3 hours.
- **Cologne:** It has very light concentrations, 2 or 3%, and lasts up to 2 hours.

Families of Perfumes

Also, there are families of fragrances according to their composition:

- **Hesperide:** Citrus, neroli, petit grain, orange blossom.
- **Floral:** Rose, jasmine, ylang ylang, tuberose, narcissus.
- **Cyprus:** Citrus, floral or green with moss and ambergris.
- **Orientals:** Civet, sandalwood, vanilla.
- **Bracken:** Lavender, oak moss.
- **Green:** Herbs, ferns, mosses and citrus.
- **Tobacco - Leather:** With woods, spices and animal background.
- **Lavender:** Lavender, ferns.
- **Wood:** Sandalwood, cedar, cypress, patchouli, vetiver.
- **Aldehydic:** Modern. Synthetics.

These families are also divided into subfamilies (soliflore, woody, bouquet, aldehydic, green, fruity, marine, cyprus, leather). To show an example, Ô de Lancôme (Lancôme, 1969), is a perfume belonging to the Hesperide or Citrus family, with high notes of bergamot, citrus and basil, medium notes of rose, ylang ylang, jasmine and carnation, and low notes of vetiver, sandalwood, oak moss and ambergris.

Raw Materials in the Perfume Industry

Raw materials for perfumes may come from different parts of the plant, such as flowers, leaves, shoots, fruits, resins, roots, rhizomes, bulbs, barks, seeds and wood.

Flowers: These are the principal source of aromas, from such species as jasmine, roses, fragrant olive (*Osmanthus fragrans*), rose (*Rosa damascena*), mimosa (*Acacia decurrens*), orange blossom or neroli (*Citrus aurantium*), ylang-ylang (*Cananga odorata*), blackcurrant buds (*Ribes nigrum*), cloves (*Eugenia caryophyllata*), hyacinth (*Hyacynthus orientalis*), jasmine (*Jasminum grandiflorum*), narcissus (*Narcissus poeticus*) and nard (*Polianthes tuberosa*), among others.

Fruits: Fresh fruits like apples, strawberries, or cherries, do not give the required aroma when extracted, so they are synthesised. Some exceptions are Verbena exotica (*Litsea cubeba*), vanilla (*Vanilla planifolia*), and juniper berries (*Juniperus communis*). The most frequently used fruits concentrate aroma in their peel, for example oranges (*Citrus aurantium*), lemons (*Citrus limon*), grapefruit (*Citrus paradisi*), and bergamot (*Citrus bergamia*), the latter being famous for aromatizing earl grey tea. From citrus three different types of aromas can be obtained, floral (called also neroli) from the buds, green from the leaves (called petit grain), and fruity from the fruit peel.

Leaves and Shoots: We should highlight patchouli, sage, rosemary, and citrus leaf oil (petit grain). Leaves are valued because of their "green" aroma, as with tomato leaves. Thyme (*Thymus* spp.), rosemary (*Rosmarinus officinalis*), sage (*Salvia officinalis*), mints (*Mentha viridis, M. spicata, M. piperita*) and pennyroyal (*Mentha pulegium*), patchouli oil (*Pogostemon cablin*) and citronella oil (*Cymbopogon winterianus*).

Resins: They have been highly valued since ancient times, incense being a good example. They are very aromatic and have antiseptic qualities. Many civilisations and cultures have used them as medicines or as condiments for food. Some resins used in perfume-making include laudanum (*Cistus ladanifer*), incense (*Calocedrus decurrens*), myrrh (*Commiphora myrrha*), Peru balm (*Myroxylon pereirae*), and tolu balm (*Myroxylon toluiferum*). Pine and fir resins are a valued source of terpenes, which are used in the organic synthesis of other synthetic aromatic substances.

Roots, Rhyzomes, and Bulbs: These sub-soil parts of the plant are also used in perfume, for example the sassafras root (*Sassafras albidum*) contains safrol, which is used in the synthesis of aromatic compounds, vetiver (*Chrysopogon zizanioides*), curcuma (*Curcuma longa*), ginger (*Zingiber officinale*) rhyzomes and orris root (*Iris germanica*).

Seeds: For example, coriander (*Anethum graveolens*), aniseed (*Pimpinella anisum*), nutmeg (*Myristica fragans*), cardamom seeds (*Elettaria cardamomum*), Angelica (*Angelica archangelica*), tonka beans (*Dipteryx odorata*).

Barks and Wood: Oils and distillations obtained from wood are basic to perfume-making. They provide the perfume's base. For example, sandalwood (*Santalum album*), birch (*Betula pendula*), jacaranda (*Jacaranda mimosifolia*), cedar (*Cedrus atlantica*), juniper (*Juniperus communis*), pine (*Pinus* spp), camphor (*Cinnamomum camphora*), fir balsam (*Abies balsamea*) and thuya (*Thuya occidentalis*). A typical example of bark is cinnamon (*Cinnamomum zeylanicum*).

Other natural sources are lichens, algae and animal products. The latter do not belong to this chapter's subject, but they are mentioned for their interest and importance. Most of them are usually produced synthetically nowadays. Oak Moss is obtained from species of lichens such as *Ramelina calicaris, Pseudoevernia furfuracea* and *Evernia prunastri.* They contain oleoresins obtained by solvent extraction. They have an intense aroma, but also act as fixatives, or substances that make the perfume last longer. Sargasso algae (*Fucus vesiculosus)*, is not so extensively used because of the high price as compared to its synthetic equivalents. Animal products used in perfumery are honeycomb, civet, ambergris, musk and whale sperm.

Other Industrial Uses of Essential Oils

Another important sector in terms of the amount of raw materials it uses and the market it creates is industrial perfume-making. This industry manufactures all sorts of products such as detergents, air-fresheners, floor-cleaners, soaps, washing-up liquids, toilet preparations. Some of the essential oils used are oils of lemon, bergamot, cedar, cypress, citronella, juniper, lavender blossoms, eucalyptus, geranium, jasmine, lavender, lilac, marjoram, pine, rosemary, salvia, sandalwood and thyme.

Industrial perfumery is used for detergents and soaps, air fresheners, deodorants, liniments, cleaning and polishing preparations, to hide the unpleasant odour of some industrial products such as rubber, plastics and paints, to provide a pleasant smell for toys, in textiles, for odour masking mordant treatments before and after dyeing, in stationery, fragrances for impregnating notebooks, cards, toilet paper, facial tissues.

Camphor wood oil (*Cinammomum camphora*), besides perfumery, is employed for the production of celluloid and various nitrocellulose compounds. Cedar wood oil (*Juniperus virginiana*), is employed for perfumery, but also in soaps, deodorants, liniments, cleaning and polishing preparations and as a cheaper adulterant of expensive sandalwood and geranium oils. It has insecticidal properties and is used as a moth repellent and in fly sprays.

CONFLICT OF INTEREST

The authors declare no conflict of interest, financial or otherwise.

ACKNOWLEDGEMENTS

Declared none.

REFERENCES

[1] Fatima A, Alok S, Agarwal P, Singh PP, Verma A. Benefits of herbal extracts in cosmetics: a review. IJPSR 2013; 4(10): 3746-60.

[2] Ribeiro AS, Estanqueiro M, Oliveira MB, Sousa Lobo JM. Main Benefits and Applicability of Plant Extracts in Skin Care Products. Cosmetics 2015; 2: 48-65. [http://dx.doi.org/10.3390/cosmetics2020048]

[3] Mantle D, Gok MA, Lennard TW. Adverse and beneficial effects of plant extracts on skin and skin disorders. Adverse Drug React Toxicol Rev 2001; 20(2): 89-103. [PMID: 11482001]

[4] Council of Europe. Plants in Cosmetics Plants and Plant Preparations Used as Ingredients for Cosmetic Products. Council of Europe Publishing 2001; Vol. II.

[5] Council of Europe. Plant Preparations Used as Ingredients of Cosmetic Products. Council of Europe Press 1998.

[6] Gupta M, Biswas TK, Saha S, Debnath PK. Therapeutic Utilization of Secretory Products of Some Indian Medicinal Plants - A Review. Indian J Tradit Knowl 2006; 5(4): 569-75.

[7] Gediya SK, Mistry RB, Patel UK, Blessy M, Jain HN. Herbal plants used as cosmetics. J Nat Prod Plant Resour 2011; 1(1): 24-32.

[8] Zheng G, Sayama K, Okubo T, Juneja LR, Oguni I. Anti-obesity effects of three major components of green tea, catechins, caffeine and theanine, in mice. In Vivo 2004; 18(1): 55-62. [PMID: 15011752]

[9] Zhu W, Gao J. The use of botanical extracts as topical skin-lightening agents for the improvement of skin pigmentation disorders. J Investig Dermatol Symp Proc 2008; 13(1): 20-4. [http://dx.doi.org/10.1038/jidsymp.2008.8] [PMID: 18369335]

[10] Benaiges A, Guillén P. Botanical Extracts.Analysis of Cosmetic Products. Elsevier 2007. [http://dx.doi.org/10.1016/B978-044452260-3/50044-9]

[11] Craciunescu O, Constantin D, Gaspar A, Toma L, Utoiu E, Moldovan L. Evaluation of antioxidant and cytoprotective activities of *Arnica montana* L. and *Artemisia absinthium* L. ethanolic extracts. Chem Cent J 2012; 6(1): 97. [http://dx.doi.org/10.1186/1752-153X-6-97] [PMID: 22958433]

[12] Cummings M, Reinhardt J, Lockhart L. Cosmet Toiletries 2000; 115: 73.

[13] Moreno Gimenez JC, Bueno J, Navas J, Camacho F. Tratamiento de las úlceras cutáneas con aceite de rosa mosqueta. Med Cut ILA 1990; XVIII: 63-6.

[14] González-Molina E, Domínguez-Perles R, Moreno DA, García-Viguera C. Natural bioactive compounds of Citrus limon for food and health. J Pharm Biomed Anal 2010; 51(2): 327-45. [http://dx.doi.org/10.1016/j.jpba.2009.07.027] [PMID: 19748198]

[15] Dulal SR, Sheikh H, Taher MA, Rahaman MS, Rahman Z, Malek MA. Formulation and finding out the efficacy of the herbal hair oil over simple coconut oil (purified) - a formulation and clinical study in Bangladesh. IJPSR 2014; 5(5): 1801-5.

[16] Krishnaveni M, Mirunalini S. Therapeutic potential of *Phyllanthus emblica* (amla): the ayurvedic wonder. J Basic Clin Physiol Pharmacol 2010; 21(1): 93-105. [PMID: 20506691]

[17] Gohil KJ, Patel JA, Gajjar AK. Pharmacological Review on *Centella asiatica:* A Potential Herbal Cure-all. Indian J Pharm Sci 2010; 72(5): 546-56. [http://dx.doi.org/10.4103/0250-474X.78519] [PMID: 21694984]

[18] Ahmed F, Urooj A. Traditional uses, medicinal properties, and phytopharmacology of *Ficus racemosa:* a review. Pharm Biol 2010; 48(6): 672-81. [http://dx.doi.org/10.3109/13880200903241861] [PMID: 20645741]

[19] al-Sereiti MR, Abu-Amer KM, Sen P. Pharmacology of rosemary (*Rosmarinus officinalis* Linn.) and its therapeutic potentials. Indian J Exp Biol 1999; 37(2): 124-30. [PMID: 10641130]

[20] Jarić S, Popović Z, Macukanović-Jocić M, *et al.* An ethnobotanical study on the usage of wild medicinal herbs from Kopaonik Mountain (Central Serbia). J Ethnopharmacol 2007; 111(1): 160-75. [http://dx.doi.org/10.1016/j.jep.2006.11.007] [PMID: 17145148]

[21] Makhija IK. LawsoniaInermis - From Traditional Use to Scientific Assessment. PharmaTutor Pharmacy Infopedia 2010; p. 113.

[22] Bozin B, Mimica-Dukic N, Samojlik I, Jovin E. Antimicrobial and antioxidant properties of rosemary and sage (Rosmarinus officinalis L. and Salvia officinalis L., Lamiaceae) essential oils. J Agric Food Chem 2007; 55(19): 7879-85. [http://dx.doi.org/10.1021/jf0715323] [PMID: 17708648]

[23] Gaur RD. Traditional dye yielding plants of Uttarakhand, India. Nat Prod Radiance 2008; 7(2): 2734-9.

[24] Chandra Shekar BR, Nagarajappa R, Suma S, Thakur R. Herbal extracts in oral health care - A review of the current scenario and its future needs. Pharmacogn Rev 2015; 9(18): 87-92. [http://dx.doi.org/10.4103/0973-7847.162101] [PMID: 26392704]

[25] Kumar G, Jalaluddin M, Rout P, Mohanty R, Dileep CL. Emerging trends of herbal care in dentistry. J Clin Diagn Res 2013; 7(8): 1827-9. [PMID: 24086929]

[26] Lupo MP. Antioxidants and vitamins in cosmetics. Clin Dermatol 2001; 19(4): 467-73. [http://dx.doi.org/10.1016/S0738-081X(01)00188-2] [PMID: 11535390]

[27] Herman A, Herman AP, Domagalska BW, Mlynarczyk A. Essential Oils and Herbal Extracts as Antimicrobial Agents in Cosmetic Emulsion. Indian Journal of Microbiology 53(2): 232-7. [http://dx.doi.org/10.1007/s12088-012-0329-0]

[28] Rivera Núñez D, Odón de Castro C. Las plantas, las esencias y los perfumes Introducción al conocimiento de sus tradiciones, cultivo y aprovechamiento en Murcia. España: Ayuntamiento de Murcia 1995.

CHAPTER 4

Industrial Uses of MAPs: Food Processing Industry

M. Paz Arraiza[1,*], Carlos Calderón-Guerrero[1], Silvia C. Guillén[2] and **Miguel A. Sarmiento[3]**

[1] *Department of Forestry and Environmental Engineering and Management, Technical University of Madrid, Spain*

[2] *Ecology Department, Agronomy Faculty, National University of Tucuman, Argentina*

[3] *Faculty of Forestry, National University of Santiago del Estero, Argentina*

Abstract: Every day the food industry uses more natural products such as spices, flavourings and food preservatives. Many aromatic and medicinal plants have antioxidant activity, germicidal and fungicidal actions which make them useful as food preservatives, attributed to volatile constituents such as phenols among other components. Precooked products gradually replace home cooked food and their preparation involves antioxidants, preservatives, flavourings and colourings of natural origin. This chapter describes the plants that are used in the manufacture of food preservatives, additives and supplements and those that are used for their organoleptic properties, such as spices and herbs.

Keywords: Condiments, Flavour, Food additives, Functional foodstuffs, Herbs, Liqueurs, Natural additives, Natural sweeteners, Preservatives, Spices.

INTRODUCTION

Every day the food industry uses more natural products such as spices, flavourings and food preservatives. Many aromatic and medicinal plants have properties such as antimicrobial, fungicide or antioxidants, attributed to constituents such as phenols among other, which make them useful as food preservatives. Precooked products gradually replace home cooked food and their preparation involves antioxidants, preservatives, flavourings and colourings of natural origin. This chapter describes the plants that are used in the manufacture of food preservatives, additives and supplements and those that are used for their organoleptic properties, such as spices and herbs.

* **Corresponding author María Paz Arraiza:** Department of Forestry and Environmental Engineering and Management, Technical University of Madrid, Madrid, Spain; Tel/Fax: +34 91 336 64 08; Email: paz.arraiza@upm.es

M. Paz Arraiza, A. González-Coloma, J. Burillo and C. Calderón-Guerrero (Eds.)

THE SPICE AND CONDIMENT INDUSTRY

Spices and condiments are vegetable substances which are used in small quantities to flavour and preserve food. Spices come from dried aromatic roots, bark, buds, seeds, and berries and other fruits. European legislation [1, 2] defines spices and aromatic condiments as *plants or parts of plants, either fresh or dried, whole, chopped or ground, whose characteristic colour, taste or smell cause them to be used to prepare foodstuffs or drinks, with the aim of incorporating said characteristics so as to make products tastier or more appetizing thus ensuring a better use of the same.*

They stimulate the appetite by increasing the taste and smell of dishes, preserve foodstuffs and may be used instead of salt as a condiment when preparing food. Many of them, such as basil, laurel, parsley, juniper, aniseed, fennel, cardamom and ginger, have been studied for antioxidant properties [3].

The spice and condiment sector is a traditional one in which small and medium-sized business coexists with multinational companies. According to CBI Market Information Database [4] some facts regarding spices in Europe are that a) internationalization promotes an increase in consumption of ethnic foods; b) the use of spices increases because they are natural ingredients; c) a healthier lifestyle promotes the consumption of products such as herbal teas; d) the market for organic food is increasing, even though the market for organic spices and herbs is still quite small.

Spices are used for meat and fish processing, canned food, sauces, soups, bakery and other pre-prepared food. The meat industry is the largest user. Mixes of spices and herbs are used in bakery, butchery, catering products, frozen meals, meat factories, oven-ready meals, salad dressings, sauces, snacks and soups. Spices and herbs derivative (extracts, oleoresins and essential oils) are used for flavouring. The beverage sector uses anise, badian and juniper in alcoholic drinks and liqueurs, and ginger in soft drinks. Herbal drinks, particularly herbal teas and infusions, are also becoming more popular [5].

According to CBI, opportunities for this sector are found in the fields of ready-t--use food (pizzas, sauces and others), health-food sector (organic spices, herbs and teas) and new spices and herbs mixes (pimento, chilly, allspice, etc.).

Spices and herbs content phenolics, which acts as antioxidant substances by terminating free radicals and quelating metal ions catalyzing lipid peroxidation [6], thus inhibiting food rancidness.

Herbs

Herbs are the leafy parts of annual plants used for food and medicine. They can be used fresh or dried. They are usually used in larger amounts than spices. Herbs of commercial interest are basil (*Ocimum basilicum*), bay leaves (*Laurus nobilis*), celery leaves (*Apium graveolens*), chives (*Allium schoenoprasum*), dill (*Anethum graveolens*), chervil (*Anthriscus cerefolium*), marjoram (*Origanum majorana*), mint (*Mentha* spp.), oregano (*Origanum vulgare*), parsley (*Petroselinum crispum*), rosemary (*Rosmarinus officinalis*), sage (*Salvia officinalis*), tarragon (*Artemisia dracunculus*), thyme (*Thymus vulgaris*) and water cress (*Nasturtium officinale*), France, Italy and Greece being important producers but low consumers of dried herbs [4].

Basil

Ocimum basilicum (Labiatae). Basil leaves are normally used fresh. There are many different types of basil. It contains methylchavicol, linalool, 1,8 - cineole, eugenol, and saponins.

Bay Leaves

Laurus nobilis (Lauraceae). Bay leaves major volatile component is 1,8 – cineole (23%), linalool (12%), α terpinyl acetate (11%) and methyl eugenol (8%) [5, 15]. It contains phenolic acids such as vainillic acid, caffeic acid and ferulic acid [6].

Celery Leaves

Apium graveolens (Umbelliferae). Its main volatile components are limonene, phellandral and a selimene [7]. It contains phenolic acids (caffeic acid, *p*-coumaric acid, and ferulic acid), and flavonoids (apigenin, luteolin, and kaempferol) [8].

Chervil

Anthriscus cerefolium (Umbelliferae). It is a delicate annual herb. The active constituents of chervil are volatile oil, whose main components are estragole (methylchavicol), flavonoids and coumarins. Chervil is used as a fine herb, known as "gourmet parsley" [9, 10].

Chives

Allium schoenoprasum (Liliaceae). It has antimicrobial and antifungal properties. Its extracts present antioxidant activity [10]. Its essential oil is rich in sulphur compounds (16.8%). The most important constituents are dipropyl disulfide, methyl pentyl disulfide, pentyl hydrodisulfide and (cis and *trans*) 3,5-diethy--1,2,4-trithiolane [11, 12].

Dill

Anethum graveolens (Umbelliferae). Dill essential oil is rich in carvone, α-phellandrene, limonene, myristicin, eugenol, and anethol, among others, which give it its characteristic odour and carminative and digestive properties. It is also rich in coumarins, flavonoids, and caffeic and chlorogenic acids, phenolic acids and steroids [13].

Marjoram

Origanum majorana (Labiatae). It is a carminative plant. Its main volatile components are carvacrol, eugenol, e and t sabinene hydrates, linalyl-acetate, myrcene, trans-anethole and terpinen 4 ol [5]. It contains antioxidants such as phenolic acids (ascorbic, caffeic, vainillic, oleanolic, rosmarinic and ursolic acids), beta carotene, beta sitosterol, hydroquinone and tannin [6, 14].

Mint

Mentha spp. (Labiatae). There are many species of mint. The most commonly used variety is peppermint (*Mentha piperita*). It contains phenolics (gallic, protochatechuic, vainillic, p-coumarinic and ferulic acids) [6]. It contains menthol, which produces a sense of freshness. Menthol is used to aromatize and as an antiseptic.

Oregano

Origanum vulgare (Labiatae). This is a wild plant which originated in the Mediterranean basin. It is normally used dried. The whole plant is rich in essential oil which contains thymol and carvacrol, camphene, gamma-terpinene, linalyl-acetate, myrcene, terpinen-4-ol, thymol [14], biocidal substances used for the preservation of food products, especially meat and cured meat. It also contains flavonoids and phenolic acids (ursolic acid, p-hydroxibenzaldehyde, p-hydroxy benzoic acid, p-coumaric acid and ferulic acid) [6].

Parsley

Petroselinum crispum (Umbelliferae). Its main volatile component is apiol [5]. It also contains phenolic acids (protocatechuic acid, gallic acid, caffeic acid, p-coumaric acid and ferulic acid) [6].

Rosemary

Rosmarinus officinalis (Labiatae). Among other substances, the essence is made up of camphor. The flower cavity contains caffeic and rosmarinic acids. It also contains flavonoids and antioxidants such as apigenin, beta-carotene, beta-sitosterol, camphene, carnosol, carvacrol, gamma-terpinene, hesperidin, hispidulin, isorosmanol, luteolin, myrcene, rosmadial, rosmanol, rosmaridiphenol, rosmariquinone, squalene, tannin, terpinen-4-ol, thymol, trans-anethole [14], and ascorbic, caffeic, carnosic, chlorogenic, labiatic, oleanolic, rosmarinic and ursolic acids [6, 14].

Sage

Salvia officinalis (Labiatae). Sage has a strong taste, which is spicy and aromatic, with a subtle hint of camphor, so it should be used in moderation. Its essential oil is rich in thujone, cineol, and borneol. Tannic matter, and bitter substances, resins, starch, albumenoids, phosphoric acid, and traces of asparagine have been found in the roots. It may also contain alanine, apigenin, beta-carotene, beta-sitosterol, camphene, carnosol, catechin, chrysoeriol, gamma-terpinene, hispidulin, linalyl-acetate, myrcene, rosmanol, stigmasterol, tannin, terpinen-4-ol, thymol, uvaol [14], and ascorbic, caffeic, carnosic, carnosolic, ferulic, fumaric, gallic, labiatic, oleanolic, palmitic, p-coumaric, p-hydroxibenzaldehyde, rosmarinic, salicylic, ursolic and vanillic acids [6, 14].

Tarragon

Artemisia dracunculus (Asteraceae). It contains an essential oil that is rich in estragol, felandrene, and ocymene, and also has high hydroxycoumarin content.

Thyme

Thymus vulgaris (Labiatae). This is one of the most typical Mediterranean spices. Its taste and smell is spicy and reminiscent of oregano. Traditionally, it was recommended as a digestive aid and to relieve coughs, sore throats by mixing thyme with honey. The essence of thyme varies greatly depending on geographical location and altitude, or on season of collection. Its essence is basically made up of thymol. Depending on the chemotype the essential oil contains 4-terpineol, anethole, camphene, carvacrol, eugenol, isoeugenol, linalyl-acetate, myrcene and gamma-terpinene. Its essence gives the plant antiseptic and anti-fungal properties, which are enhanced by phenolic acids such as caffeic, rosmarinic, chlorogenic, isochlorogenic, lauric, coumaric, ferulic, labiatic, gallic, myristic, oleanolic p-hydroxy-benzoic ursolic and vanillic acids. Apart from this, this aromatic plant contains flavonoids and alanine plant, apigenin, beta-carotene, chrysoeriol, eriodictyol, luteolin, methionine, naringenin, tannin and tryptophan [14].

Spices

Spices come from any part of a plant such as the roots, bark, flowers, fruit, and seeds. They are usually native to tropical climates. They are used dried and have traditionally been used as a preservative, as they have antimicrobial, antiseptic, and antibacterial properties. Spices of commercial interest are anise (*Pimpinella anisum*), caraway (*Carum carvi*), cardamom (*Elettaria cardammomum*), cinnamon (*Cinammomun zeylanicum*), cloves (*Eugenia caryophilla*), coriander (*Coriandrum sativum*), cumin (*Cuminum cyminum*), fennel seeds (*Foeniculum vulgare*), fenugreek (*Trigonella foenum-graecum*), ginger (*Zingiber officinale*), juniper berries (*Juniperus communis*), mace and nutmeg (*Myristica fragans*), paprika (chillies and cayenne pepper, *Capsicum annuum*), pepper (*Piper nigrum*), pimento (*Pimenta dioica*), saffron (*Crocus sativus*), starred aniseed or badian seeds (*Illicium verum*), turmeric (*Curcuma longa*) and vanilla (*Vanilla platifolia*).

Allspice

Pimenta dioica, (Myrtaceae). It contains eugenol and caryophyllene [5]. It is stimulant, carminative and digestive spice. Its name makes reference to the fact that it combines the flavours and aromas of cinnamon, clove and nutmeg.

Aniseed

Pimpinella anisum (Umbelliferae). Aniseed is the schizocarpic fruit of the annual anise plant [5]. Aniseed essential oil is rich in anethole (80%), anisaldehyde, anisketone and methyl chavicol. It contains other ingredients such as furanocoumarins, flavonoids, phenylpropanoids, sterols, proteins [15], glycosides, fixed oils, fatty oils (30%) and choline [14].

Caraway

Carum carvi (Apiaceae). Its main volatile components are d-carvone camphene plant 0–30, carvacrol plant, gamma-terpinene fruit 12–14,160, gamma-terpinene plant 0–270 myrcene and terpinen-4-ol [5]. Other active compounds with antioxidant activity are Beta-carotene, lauric-acid, myristic-acid, myristicin, palmitic-acid, quercetin and tannin [13].

Cardamom

Elettaria cardammomum (Zingiberaceae) is one of the most expensive spices after saffron and vanilla, used in food preparations, perfumery, spices, medicines and beverages [15]. Its main volatile components are α-terpinyl acetate (30%), 1-8 cineole (36%) and limonene (11%), combination which provides its mild spicy flavour [5, 15].

Cinnamon

Cinnamomun zeylanicum (Lauraceae). Cinnamon is extracted from the inner bark of the cinnamon tree. It is one of the most widely used spices in the world. Cinnamon bark contains essential oil, cinnamic aldehyde, tannins, terpenes, calcium oxalate, starch, and mucilage. It contains antioxidant phenolics (vainillic, caffeic and ferulic acids) [6]. The effect of all these substances together is to give it digestive properties, making it a stomach tonic and an appetizer.

Clove

Syzygium aromaticum (Mirtaceae). Cloves are the unopened flowers (buds) of a tropical tree which are dried in the sun. Aromatic cloves are found in the cuisine of Europe, Africa, Arab countries, and the Americas. Cloves contain an abundance of essential oil, are rich in eugenol, acetyl eugenol, caryophyllene, pinene, caryophylline, and methyle salycilate.

Coriander

Coriandrum sativum. (Umbelliferae). Originally from the Mediterranean region, coriander is grown all over the world nowadays. Its fruit contains an essential oil, linalool, whose active ingredient gives it eupeptic and carminative properties. The seeds contain 0.5 - 1% essential oil rich in carvone, geraniol, limonene, borneol, camphor, elemol, myrcene, gamma-terpinene, terpinen-4-ol, trans-anethole and linalool. It also contains the flavonoids quercitin, kaempferol, rhamnetin, and apigenin, and phenolic acid compounds such as caffeic and chlorogenic acid [4]. Other antioxidant components are apigenin, ascorbic-acid, beta-carotene, beta-sitosterol, isoquercitrin, myristic-acid, myristicin, *p*-hydroxy-benzoic-acid, palmitic-acid, protocatechuic-acid, quercetin, rhamnetin, rutin, scopoletin, tannin and vanillic-acid [14].

Cumin

Cuminum cyminum (Umbelliferae). This is one of the most widely used spices in the world. Cumin is the basic ingredient of curry and it contains 10% of essential oil whose main active ingredient is cuminaldehyde (40-65%) [15]. It is an appetizer, and is both digestive and carminative.

Fennel Seeds

Foeniculum vulgare (Umbelliferae). Its flavour is similar to anisa and star anise but not so strong. The main volatile components are anethole, limonene, α pinene, fenchol and fenchone and estragole [5, 15].

Fenugreek

Trigonella foenum-graecum (Fabaceae). Fenugreek plant seeds are used in seasonings, pickles and curries. It is used as both herb and spice. The seed is made up by two parts, the testa, containing essential oil, saponin and protein, and the albumen containing mainly glucomannan. The dominant aroma is given by a hemiterpenoid g lactone, soloton [15].

Ginger

Zingiber officinale (Zingiberaceae). It is used in foods and beverages. It is used fresh but its preserved and dried products are the most widely used. Its volatile components include monoterpenes, which are the main contributors to its typical aroma, such as 1-8 cineol, citral, neral, geraniol, linalool, alfa terpineol, D borneol and bornyl acetate [15]. It also contains alanine, ascorbic acid, beta carotene, beta sitosterol, caffeic acid, capsaicin, chlorogenic acid plant, curcumin, lauric acid, methionine, myristic acid, p-coumaric acid, palmitic acid, tryptophan [14].

Nutmeg

Myristica fragrans (Myristicaceae). Seed or drupe of nutmeg. Its components include several terpenes (eugenol, borneol, geraniol, saffrol, camphene gamma-terpinene, isoeugenol, methyl-eugenol, myrcene, terpinen-4-ol), and 3-4% myristicine, which is toxic at high dosages. Other antioxidant components are lauric-acid, myristic-acid, oleanolic acid and palmitic acid [14]. Mace is the dried reticulated aril of nutmeg, containing alfa pinene, sabinene, and terpinen 4 ol [5].

Paprika

Capsicum annuum (Solanaceae). Paprika is the result of grinding the dried fruit of peppers. Christopher Columbus claimed to have discovered it on his voyages to America. It is used as a condiment and also to preserve food, especially meat. At an industrial level, it is commercialized as oleoresin. Paprika's main active ingredient is capsaicin, whose chemical name is methylenonenic acid vanililamide. It also contains antioxidants such as alanine, alpha-tocopherol, ascorbic-acid, beta-carotene, caffeic acid, campesterol, camphene, capsanthin, chlorogenic-acid, eugenol, gamma-terpinene, hesperidin, histidine, lutein, methionine, myrcene, myristic-acid, pcoumaric-acid, palmitic-acid, pentadecanoic-acid, scopoletin, stigmasterol, terpinen-4-ol, tocopherol and tryptophan [14].

Pepper

Piper nigrum (Piperaceae). Pepper grains can be green, black, or white, depending on the different stages of ripening. The main fact about the origins of pepper is a mention in ancient Indian texts. From India, trade spread pepper to Malaya and Indonesia. Later it reached the New World too. It contains two makor types of constituents, the volatile or essential oil (1-4%) with α and β pinenes, β caryophyllene, D limonene felandrene, myrcene, sabinene, and the resin, containing piperine and other secondary metabolites (piperiline, piperitine, piperinine), responsible of the pungency [15, 16].

Saffron

Crocus sativus (Liliaciae). It is called "red gold" because it is one of the most expensive spices. It is used as a dye and in perfumes, but its most important uses are culinary. Saffron contains a bitter ingredient (picrocrocin), a colouring agent (crocin), and essential oil.

Turmeric

Curcuma longa (Zingiberaceae). Its essential oil contains turmerone, zingeberene, p-cymene, 1,8 cineole [5, 15]. Depending on the part of the plant, the leaves contain linalool, the flower a terpineol, and roots and rhizome, turmenone [15]. It also contains ascorbic-acid, beta-carotene, caffeic-acid, curcumin, eugenol, *p*-coumaric-acid, protocatechuic-acid, syringic-acid and vanillic acid [14].

Vanilla

Vanilla platifolia (Orchidaceae). Vanilla beans have been used since the time of the Aztecs, and are now considered the most important flavouring in the food and perfume industries. They develop their aroma when green vanilla beans are processed to produce cured vanilla beans. They contain vanillin, vanillic acid, *p*-hydroxybenzaldehyde and *p*-hydroxybenzoic acid [17].

NATURAL FOOD ADDITIVES

Food additives are chemical substances which are added in small quantities to foodstuffs to preserve and to improve texture, taste, appearance, and smell. There is a wide variety of additives, but mainly preservatives (antioxidants and antimicrobials), colorants, sweeteners, thickeners and flavours [18, 19]. This market has become more stable over the past 5 years, as compared to synthetic equivalents. It is expected that this market will grow during the next years [19].

Natural antioxidants include polyphenols, carotenoids, ascorbic acid and tocopherol. Natural antimicrobials include polyphenols and essential oils such as carvacrol, thymol carnosic acid and capsaicin. Natural flavours include substances such as oleoresins, essential oils of mint, peppermint, menthol. Natural colours include substances such as beta-carotene, lutein, astaxanthin, capsanthin, annatto, betalains, paprika, anthocyanins, chlorophylls and curcumin. Natural thickeners include substances such as locust bean gum, guar gum, tara gum, seaweeds, algae and pectin. Natural sweeteners include steviol glycosides, glycyrrhizin and thaumatin [18, 19].

The most commercially important natural sweetener is stevia (*Stevia rebaudiana*). It contains steviol glycosides such as stevioside (5-10%), rebaudiosides (A, 2-5%; C, 1%; D, E, F 0.2%), dulcoside (0,5%) and steviolbioside (0.1%) [20].

Natural Preservatives of Plant Origin

Preservatives are natural and artificial substances used to preserve foodstuffs against the effects of micro-organisms, so as to stop them deteriorating for a certain period of time and under certain storage conditions. There are two main types, antioxidants and antimicrobials.

A good example of antioxidant is rosemary extract, E392 according to Council Regulation (EC) 1129/2011, which contains specific constituents such as carnosic acid. Other antioxidants are ferulic acid, catechin, ascorbic acid (E300), carotenoids (lycopene, E160d), and tocopherol or vitamin E [18].

Antimicrobials inhibit the growth and proliferation of microorganisms, and they include terpenes (such as carvacrol, eugenol and thymol), steroids, alkaloids and polyphenols. Another important source of antimicrobials is essential oils, such as cinnamon, clove, oregano, rosemary and thyme.

Natural Coloring Agents of Plant Origin

Coloring agents are natural or artificial substances which are used to enhance the color of foodstuffs, either because the food product has lost color during industrial treatment, or to make it look tastier and more attractive to the consumer. Some natural coloring agents are:

Anatto (E160b)

Annatto is a permitted natural food colorant, extracted from the *Bixa orellana* tree [18]. It gives a yellow – orange color used in dairy products and bakery.

Anthocyanins (E163)

They are red, purple, violet and blue pigments responsible for pigments in fruits and flower colors. The main anthocyanins are cyaniding, cyanidin, malvidin, pelargonidin, peonidin, petunidin, and their main use is in soft drinks, confectionery products and fruit preparation [20].

Chlorophyll (E140)

It is used in ice-cream, cakes, drinks, condiments, vinaigrettes, etc. It is presented in a hydro-soluble liquid, lipo-soluble liquid, or hydro-soluble powder form. It has medium-to-good stability in the presence of acids, and medium stability to light and heat. It gives a green color. Of the 5 existing chlorophylls, only a and b chlorophyll are used as food colorants.

Carminic Acid (E120)

The word carmine defines both a shade of color and a coloring agent. In fact, carmine is the name usually given to a natural red coloring agent extracted from the female cochineal, *Dactylopius coccus*, an insect that lives in the branches of cactuses, especially in the *Opuntia coccinellifera* found mainly in Peru and the Canary Islands, Spain. This coloring agent has been in use since ancient times to treat cloth, vinegar, alcohols, and meat products. More recently it is used in cosmetics. It is very expensive. Carmine gives a bright red color in an acid medium, and a violet color in an alkaline medium. This is because carmine belongs to the anthocyanin group whose pH indicators depend on the medium.

Curcumin (E100) (Curcuma longa)

It is used in ice-cream, sauces, soups, cakes, desserts, pre-cooked meals, cheeses, drinks, condiments, etc. It is used as a hydro-soluble liquid, lipo-soluble liquid or hydro-soluble powder form. It has good stability in the presence of acids, medium stability to heat, and little stability to light. It gives a yellow or yellowy-orange color.

Paprika (E160c) (Capsicum annuum)

Paprika is a mixture of two carotenoids, capsanthin and capsorubin, provides an orange to red color [20].

FUNCTIONAL FOODSTUFFS

Nowadays, a very important commercial demand for MAPs is for the so called functional foodstuffs and nutraceuticals. There are no universally accepted definitions for them, but in general they are described as foods (fortified with added ingredients or not) with health benefits beyond basic nutrition [21, 22].

The most commonly accepted definition of a functional foodstuff is: a foodstuff that is similar in appearance to a traditional foodstuff, which is eaten as part of a normal diet, and which goes beyond its nutritional value because it has been scientifically proven to reduce the risk of getting a particular illness and/ or to have physiological benefits.

They have opened up opportunities for value addition and new product development in the food and pharmaceutical industries [23], but, according to [24], most of the functional foods belong to the food sector (red grape for its antioxidant properties, lycopene from tomatoes, and broccoli for its cancer preventative properties) more than to the medicinal sector.

However, there are many examples of medicinal and aromatic plants in this sector, such as black cohosh (*Cimifuga racemosa*) [25] for menopause symptoms, *Ginkgo biloba* for memory loss [26], and tea, *Camellia sinensis*, for its antioxidants [27]. Some of these phytochemicals are carotenoids (betacarotene, licopene, lutein), glucosinolates, phytoestrogens and polyphenols (lignans, phenolic acids, tannins, flavonoids).

THE LIQUOR INDUSTRY

Aromatic and medicinal plants are used to prepare a wide range of liquors. From the XV-XIX centuries, these were used for medicinal purposes. Among the many plant species used there are barks such as cinnamon bark (*Cinnamomum zeylanicum*); flowers such as clove buds (*Syzygium aromaticum*) and vanilla pods (*Vanilla planifolia*); fruits such as juniper berries (*Juniperus communis*); leaves such as fennel (*Foeniculum vulgare*), hyssop (*Hyssopus officinalis*), marjoram (*Origanum majorana*), lemon balm (*Melissa officinalis*), mint (*Mentha piperita*), oregano (*Origanum vulgaris*) and thyme (*Thymus vulgaris*); roots such as Angelica (*Angelica archangelica*), gentian (*Gentiana lutea*) and turmeric (*Curcuma longa*); seeds such as aniseed (*Pimpinela anisum*), cardamom (*Elettaria cardammomum*), coriander (*Coriandrum sativum*) and starred aniseed (*Illicium verum*).

Production Methods

There are three basic methods of elaborating liquors, two cold, and one using heat. The method used depends on the plants, the aroma and the taste. Fruits are best extracted cold, whereas seeds, peels, and flowers are usually extracted hot. The parts of the plant used are seeds (almonds), fruits or berries (juniper), peels (citrus), roots (licorice), flowers (orange flowers), and leaves (mint). Natural concentrates are also used. Liquors may be classified according to their main components.

Liquors Made of Herbs and Spices

They contain herbs such as basil, rosemary, salvia, mint. Examples include Benedictine® (French liqueur made of more than 27 plants, it was used as a remedy for malaria), Chartreuse® (it contains more than 130 plants) and Drambuie® (it contains licorice, herbs and whisky).

Liquors Made from Seeds

They contain seeds like aniseed, hazelnuts, or almonds. Examples include Anise, very popular in Spain, France and Greece. It contains anise seeds. Distilled it is between 40-60% proof. Ouzo® is a Greek liquor which contains aniseed seeds, licorice, mint, fennel, and hazelnuts. Pernod® is French liquor which is a variety of traditional absinthe. It is an herbal elixir containing 15 different herbs, among which is absinth (*Artemisia absinthium*). In Spain it is distilled with cumin seeds (*Ciminum cuminum*).

Gin

In 1650, Franciscus de la Boe, a Dutch doctor and Professor at Leyden University, discovered the diuretic and hepatic qualities of the juniper berry, and developed a medicinal tonic using liquors and said berries. He named it Geniévre (juniper in French). It also contains coriander, angelica roots, liquorice, cassia bark (*Cinammomum cassia*), and orange and lemon peel. The Bombay® variety contains Moroccan coriander seeds, Chinese liquorice, Spanish almonds, angelica from Saxony, iris roots (*Iris germanica*), Italian juniper, cassia bark from Indochina, Javanese pepper (*Piper cubeba*), and paradise grains (*Aframomum melegueta)* from Africa.

Vermouth

The word “vermouth” is the French equivalent to the German “wermut” or English “wormwood” (absinthe, *Artemisa absinthium*). It is a fortified wine, with herbs and spices, the main ones being angelica, aloe juice, cinnamon, cloves, coriander, gentian, hyssop, lemon balm, marjoram, sweet and sour orange peel, quinine, salvia, thyme and absinthe.

CONFLICT OF INTEREST

The authors declare no conflict of interest, financial or otherwise.

ACKNOWLEDGEMENTS

Declared none.

REFERENCES

[1] Lachenmeier DW. Wormwood (Artemisia absinthium L.)--a curious plant with both neurotoxic and neuroprotective properties? J Ethnopharmacol 2010; 131(1): 224-7. [http://dx.doi.org/10.1016/j.jep.2010.05.062] [PMID: 20542104]

[2] Bora KS, Sharma A. Phytochemical and pharmacological potential of *Artemisia absinthium* Linn. and *Artemisia asiatica* Nakai: A Review. J Pharm Res 2010; 3: 325-8.

[3] Hinneburg I, Dorman HD, Hiltunen R. Antioxidant activities of extracts from selected culinary herbs and spices. Food Chem 2006; 97(1): 122-9. [http://dx.doi.org/10.1016/j.foodchem.2005.03.028]

[4] Database CMI. CBI market survey. CBI 2009.

[5] Peter KV, Ed. Handbook of herbs and spices. Woodhead publishing 2006; Vol. 6. [http://dx.doi.org/10.1533/9781845691717]

[6] Muchuweti M, Kativu E, Mupure CH, Chidewe C, Ndhlala AR, Benchura MA. Phenolic composition and antioxidant properties of some spices. Am J Food Technol 2007; 2(5): 414-20. [http://dx.doi.org/10.3923/ajft.2007.414.420]

[7] Alves-Silva JM, dos Santos SM, Pintado ME, Pérez-Álvarez JA, Fernández-López J, Viuda-Martos M. Chemical composition and in vitro antimicrobial, antifungal and antioxidant properties of essential oils obtained from some herbs widely used in Portugal. Food Control 2013; 32(2): 371-8. [http://dx.doi.org/10.1016/j.foodcont.2012.12.022]

[8] Yao Y, Sang W, Zhou M, Ren G. Phenolic composition and antioxidant activities of 11 celery cultivars. J Food Sci 2010; 75(1): C9-C13. [http://dx.doi.org/10.1111/j.1750-3841.2009.01392.x] [PMID: 20492156]

[9] Vyas A, Shukla SS, Pandey R, Jain V, Joshi V, Gidwani B. Chervil: A multifunctional miraculous nutritional herb. Asian J Plant Sci 2012; 11(4): 163. [http://dx.doi.org/10.3923/ajps.2012.163.171]

[10] Thorn B. Chervil: herb with verve leaves chefs enthralled with an alternative to parsley. Nation's Restaurant News 2000; 34(27)

[11] Štajner D, Popović BM, Calić-Dragosavac D, Malenčić D, Zdravković-Korać S. Comparative study on *Allium schoenoprasum* cultivated plant and *Allium schoenoprasum* tissue culture organs antioxidant status. Phytother Res 2011; 25(11): 1618-22. [http://dx.doi.org/10.1002/ptr.3394] [PMID: 21438046]

[12] Hashimoto S, Miyazawa M, Kameoka H. Volatile flavor components of chive (*Allium schoenoprasum* L.). J Food Sci 1983; 48(6): 1858-9. [http://dx.doi.org/10.1111/j.1365-2621.1983.tb05101.x]

[13] Jana S, Shekhawat GS. *Anethum graveolens*: An Indian traditional medicinal herb and spice. Pharmacogn Rev 2010; 4(8): 179-84.

[http://dx.doi.org/10.4103/0973-7847.70915] [PMID: 22228959]

[14] Suhaj M. Spice antioxidants isolation and their antiradical activity: a review. J Food Compos Anal 2006; 19(6): 531-7.
[http://dx.doi.org/10.1016/j.jfca.2004.11.005]

[15] Parthasarathy, VA: CABI, ProQuest Ebook Central 2008.

[16] Ravindran PN. Black pepper: Piper nigrum series: Medicinal and aromatic plants. Kerala, India: Centre for Medicinal Plant Research 2000.

[17] Pérez-Silva A, Odoux E, Brat P, Ribeyre F, Rodriguez-Jimenes G, Robles-Olvera V, *et al.* GC–MS and GC–olfactometry analysis of aroma compounds in a representative organic aroma extract from cured vanilla (*Vanilla planifolia* G. Jackson) beans. Food Chem 2006; 99(4): 728-35.
[http://dx.doi.org/10.1016/j.foodchem.2005.08.050]

[18] Carocho M, Morales P, Ferreira IC. Natural food additives: *Quo vadis?* Trends Food Sci Technol 2015; 45(2): 284-95.
[http://dx.doi.org/10.1016/j.tifs.2015.06.007]

[19] CBI. Natural Colours, Flavours and Thickeners Ministry of foreign affairs: The Netherlands 2015.

[20] Henry BS. Natural food colours. 1996.
[http://dx.doi.org/10.1007/978-1-4615-2155-6_2]

[21] Tavarini S, Angelini LG. *Stevia rebaudiana* Bertoni as a source of bioactive compounds: the effect of harvest time, experimental site and crop age on steviol glycoside content and antioxidant properties. J Sci Food Agric 2013; 93(9): 2121-9.
[http://dx.doi.org/10.1002/jsfa.6016] [PMID: 23303701]

[22] Wildman RE. The nutritionist: food, nutrition, and optimal health. Routledge 2009.

[23] Childs NM, Poryzees GH. Foods that help prevent disease: consumer attitudes and public policy implications. Br Food J 1998; 100(9): 419-26.

[24] Lubbe A, Verpoorte R. Cultivation of medicinal and aromatic plants for specialty industrial materials. Ind Crops Prod 2011; 34(1): 785-801.
[http://dx.doi.org/10.1016/j.indcrop.2011.01.019]

[25] Shams T, Setia MS, Hemmings R, McCusker J, Sewitch M, Ciampi A. Efficacy of black cohosh-containing preparations on menopausal symptoms: a meta-analysis. Altern Ther Health Med 2010; 16(1): 36-44.
[PMID: 20085176]

[26] Kaschel R. Specific memory effects of *Ginkgo biloba* extract EGb 761 in middle-aged healthy volunteers. Phytomedicine 2011; 18(14): 1202-7.
[http://dx.doi.org/10.1016/j.phymed.2011.06.021] [PMID: 21802920]

[27] Nagarani B, Debnath S, Kumar SC, Bhattacharjee CB, Kumar GG. A review: Herbs used as anticancer agents. Int Res J Pharm 2011; 2(1): 20-4.

CHAPTER 5

Industrial Uses of MAPs: The Pharmaceutical Sector

M. Paz Arraiza[1,*], **Carlos Calderón-Guerrero**[1], **Silvia C. Guillén**[2] and **Miguel A. Sarmiento**[3]

[1] *Department of Forestry and Environmental Engineering and Management, Technical University of Madrid, Spain*

[2] *Ecology Department, Agronomy Faculty, National University of Tucuman, Argentina*

[3] *Faculty of Forestry, National University of Santiago del Estero, Argentina*

Abstract: Medicinal and aromatic plants play an important role in the pharmaceutical sector, both in the phytotherapy aspect, as in the strictly medicinal significance. Aromatherapy is the use of essential oils for therapeutic purposes. The absorption can be topical, oral or by inhalation. The second group includes all preventive and therapeutic use of MAP or their active ingredients. Active principles from medicinal plants are called phytopharmaceuticals, phytomedicines, herbal medicines or botanicals. Current pharmacopoeia reached more than 7,000 compounds derived from these plants. The uses of these substances include: 1) health care using natural products; 2) The newer drugs against the challenges of the twenty-first century (cancer, cholesterol, diabetes, dementia, depression, malaria, stress, *etc.*); 3) Essential oils used in aromatherapy; and 4) compounds useful in the veterinary area. The main species used in each of the previous four sections are introduced in this chapter.

Keywords: Aromatherapy, Cancer, Dementia, Depression, Diabetes, Essential oils, Fatigue, Herbal medicines, Hypercholesterolemia, Malaria, Menopause, Natural antibiotics, Natural products, Pharmaceuticals, Phytotherapy, Stress.

INTRODUCTION

The pharmaceutical industry is the most important application of medicinal and aromatic plants. This group includes all preventive and therapeutic use of MAP or their active ingredients, *i.e.*, the use of MAPs due to their pharmacological properties.

* **Corresponding author María Paz Arraiza:** Department of Forestry and Environmental Engineering and Management, Technical University of Madrid, Madrid, Spain; Tel/Fax: +34 91 336 64 08; Email: paz.arraiza@upm.es

M. Paz Arraiza, A. González-Coloma, J. Burillo and C. Calderón-Guerrero (Eds.)

According to P. J. Houghton [1] 25% of the pharmaceuticals prescribed in the developed world contain chemicals produced by flowering plants. If fungi and some animals are included, it rises to 40% [2].

Active principles form medicinal plants are called phytopharmaceuticals, phytomedicines, herbal medicines or botanicals, and it is important to understand that their medicinal properties are not based on a single chemical entity (SCE) but rather on the synergy of the multiple compounds present in the drug [1].

The pharmaceutical industry is the most important user of MAPs. We include, in this group, all therapeutic and preventive uses, *i.e.*, the use of medicinal plants for their pharmacological activity.

The pharmacological activity of medicinal plants is obviously conditioned by the presence of certain active molecules, which are able to bond with different targets in living beings, thus modifying them. Synthetic drugs do the same.

The main difference lies in the concentration of these molecules in the drug. This is generally less in the plant, which produces significant differences of action, but also important differences as far as side-effects go. Phytotherapeutic drugs have a less intensive pharmacological action, take longer to act, but also have fewer adverse effects than synthetic ones.

Some species stand out above the rest because of a specific compound responsible for the pharmacological activity. However, it is frequently the case that the effects of administering a medicinal plant are different to those of administering just the components in isolation. Therefore, there exists a possibility that the final effect may be due to synergism (enhancing of activity) or to antagonism (reducing of activity) between the different active ingredients which the plant contains.

On the other hand, some components may interfere in the pharmacological activity of the active ingredients by modifying their pharmacokinetics, increasing or reducing their bio-availability. This is why the efficiency of certain phytotherapeutic medicines has not been improved by the administration of isolated active ingredients, as in the cases of the roots of *Valeriana officinalis* (sedative) [1, 3], or the roots of the ginseng, *Panax ginseng* (stimulant) [3, 4].

The European Union recognises the importance of phytotherapy, especially in terms of treatment and prevention of minor ailments, but also for certain chronic afflictions. Phytotherapy may be used as an alternative therapy in cases where intolerance to or inefficiency of synthetic drugs are manifest, and also as a complementary treatment for serious diseases. The World Health Organization,

(WHO) [5], on its part, recommends the use of phytotherapy and takes it as the only form of therapy possible in developing countries. According to the WHO [5]:

- 75% of modern plant-derived pharmaceutical medicines are used in ways that are similar to their traditional uses.
- 25% of modern drugs used in the United States have been derived from plants.
- 75% of plants that provide active ingredients have been studied because of their use in traditional medicine.
- 75% of plant drugs show correlation between their current therapeutic use and the traditional use of the plant from which they derive.
- More than 2/3 of the world's plant species (about 35,000 are medicinal) come from developing countries.
- At least 7,000 medical compounds in the modern pharmacopoeias are derived from plants.

A total of 19 natural product based drugs were approved for marketing worldwide between 2005 and 2010, among which 7 are classified as natural products (Dronabinol/Cannabidol, Fumagillin, Ziconotide, Exenatide, Trabectedin, Romid-epsin and Capsaicin), 10 as semi-synthetic natural products, and 2 as natural products -derived drugs. The problems treated with these 7 natural product drugs are pain, parasites and cancer [6].

Some examples of plants used as medicine nowadays are artichoke (*Cynara cardunculus*), which contains sesquiterpene lactones valid for lowering cholesterol levels in blood. Black cohosh (*Cimifuga racemosa*) contains phytoestrogens useful for the treatment of menopause symptoms. Black cumin (*Nigella sativa*) can be used against coughs, pulmonary infections, asthma, influenza, allergy, hypertension and stomach ache. Purple coneflower (*Echinacea purpurea*) extracts are useful for cardiac problems. Feverfew (*Tanacetum parthenium*) can be used to treat migraine headaches. Garlic extract (*Allium cepa*) can lower total cholesterol levels. Oregano (*Origanum vulgare*) may be effective against multi-drug resistant bacteria. Pawpaw (*Asimina triloba*) can be used for insecticidal purposes (killing lice, worms). Peppermint (*Mentha piperita*) oil may have benefits for irritable bowel syndrome. Pokeweed (*Phytolacca americana*) is used in homeopathy. Saw Palmetto (*Serenoa repens*) can be used for benign prostatic hyperplasia, BPH. Snake root (*Rawolfia serpentina*) is used for high blood pressure. Spanish sage (*Salvia lavandulifolia*) may improve memory in Alzheimer's disease. St. John's wort (*Hypericum perforatum*) has been used for the treatment of depression. Valerian root (*Valeriana officinalis*) can be used to treat insomnia.

Another well-known but controversial plant that has received increasing attention for its medicinal properties is cannabis (*Cannabis sativa*). Clinical trials provide evidence for its therapeutic use [7, 8]. If the acceptance of the medicinal use of cannabis continues, there will be an increase in the demand for good quality cannabis and cannabis-based preparations in the near future.

MEDICINAL PLANTS FOR MODERN HEALTH CHALLENGES

According to the CBI [9, 10, 11], natural and healthier lifestyle increasingly demands natural healthcare options that bring an increasing demand for medicinal and aromatic plants and extracts. In this sense, the demand for organically certified and fair trade raw materials is also increasing. More reports list current trends for phytopharmaceuticals:

- Plants with active principles against cancer.
- Plants with active principles against dementia.
- Plants with active principles against diabetes.
- Plants with active principles that help relieve stress and fatigue.
- Plant substances to lower hypercholesterolemia.
- Plants with anti-malarial active principles.
- Plants containing active principles for women's health and menopause.
- Plants with natural antibiotics.

Other specific health problems for which medicinal plants are useful are obesity (*Ilex paraguayiensis, Capsicum annuum*), joints (*Curcuma longa, Boswellia serrata*) and digestive health (*Pneumus boldus, Plantago afra*), cold and flu (*Echinacea purpurea*) or arthritis (*Harpagophytum procumbens*).

Cancer, the Greatest Challenge in the 21st Century

Anticancer agents from plants currently in use belong to four main classes of compounds: vinca alkaloids, epipodophyllotoxins, taxanes, and camptothecins [12]. Vinca alkaloids, vinblastine and vincristine, were isolated from *Catharanthus roseus* (Apocynaceae) and have been used clinically for over 40 years [13]. Podophyllotoxin was isolated from the resin of *Podophyllum peltatum* (Berberidaceae) and from it the drug etoposide was approved. Paclitaxel was isolated from *Taxus brevifolia* (Taxaceae) and was clinically introduced to the market in the early 1990s. Camptothecin was isolated from *Camptotheca acuminata* (Nyssaceae). Many derivatives of all of them have been synthesized, some of which are currently in clinical use, and all of these natural products have led to important discoveries related to their mechanisms of action [12].

Other compounds that have been researched for cancer prevention and treatment are terpenes, alkaloids, polyphenols and other substances [7]. Terpenes such as limonene, *Perilla frutescens* alcohol [14], triterpens from *Ganoderma lucidum* fungus [15], terpenes found in *Zizyphus mauritana* [16]; glycosides such as ginsenosides from *Panax ginseng*; alkaloids such as ecteinascidin from *Ecteinascidia turbinata* [17], or camptothecine from *Camptotheca acuminata* [18]; polyphenols such as genistein, an isoflavone found in *Trifolium pratense* or *Glycine max*; aurones from *Uvaria hamiltonii* [19] or combrestatin from *Combretum caffrum* [20]; other substances such as curcuminoids from *Zingiber mioja* or discodermolide from *Gymnodium* spp. Another plant for cancer treatment is saw palmetto (*Serenoa repens*) for the treatment of symptomatic benign prostatic hyperplasia [21].

Dementia and Memory Loss

According to the Lubbe and Veerporte [10], examples of popular phyto-pharmaceuticals are *Gingko biloba* extract to improve cognitive function [3], ginseng (*Panax ginseng*) as general tonic and cognitive enhancer [4]. Some plants that improve cognition and concentration support are (*Centella asiatica, Withania somnifera*) [9]. There are Ayurvedic medicinal polyherbs preparations, such as Trasina or Memorin, used in Alzheimer's disease [22]. Traditional Chinese Medicine has accumulated experience of treating dementia using plant-derived medicines. A recent analysis of Traditional Chinese Medicine formulae revealed that the most common combination of herbs and drugs used for this purpose was *Rhizoma chuanxiong, Radix salvia miltiorrhizae, Radix polygalae tenuifoliae* and *Rhizoma acori tatarinowii* [23].

Diabetes

Our modern lifestyle's needs and trends are anti-diabetes plants, such as garlic (*Allium sativum*), bitter gourd (*Momordica charantia*) and prickly pear (*Opuntia ficus-indica*) [9]; Holy basil *(Ocimum album)* leaves decrease blood glucose levels in patients with diabetes mellitus in cursive [22, 24].

Curcumin is the yellow-coloured bioactive constituent of turmeric (*Curcuma longa).* Scientific literature shows that curcumin possesses anti-diabetic effects and mitigates diabetes complications. Anti-diabetic activity of curcumin may be due to its potent ability to suppress oxidative stress and inflammation. Curcumin reduces blood glucose [25].

Some natural products have been reported to activate AMPK (AMP-activated protein kinase), a cellular energy detector that helps energy balance of the cells and calorie consumption, and is particularly relevant to treatment of type 2 diabetes. These include galegine and guanidine, from *Galega officinalis*, leading to the development of the biguanides, resveratrol from red grapes, quercetin, ginsenoside from *Panax ginseng*, curcumin from *Curcuma longa*, berberine from

Coptis chinensis, epigallocatechin gallate from green tea (*Camellia sinensis*), theaflavin from black tea (*Camellia sinensis*), and hispidulin from snow lotus (*Saussurea involucrata*) [26].

Fatigue, Stress and Depression

Rosenroot, *Rhodiola rosea*, is a popular plant in traditional medical systems in Eastern Europe and Asia, stimulating the nervous system, decreasing depression, enhancing work performance, eliminating fatigue, and preventing high altitude sickness. It has been categorized as an adaptogen due to its observed ability to increase resistance to a variety of chemical, biological, and physical stressors. Its benefits include antidepressant, anticancer, cardioprotective, and central nervous system enhancement. Research also indicates great utility in asthenic conditions (decline in work performance, sleep difficulties, poor appetite, irritability, hypertension, headaches, and fatigue) developing subsequent to intense physical or intellectual strain [27]. Its main components are rosidirol and its derivate rosidirin [28].

Liquorice (*Glycirrhiza glabra*) has been reported for successful treatment of Chronic Fatigue Syndrome (CFS) due to its ability to improve adrenal function [29]. Ginseng, *Panax ginseng*, containing ginsenosides and eleutherosides, is used against fatigue and stress [30]. Ginkgo, *Ginkgo biloba*, contains ginkgolides (terpenetrilactones), flavonol glycosides and is used to treat mental fatigue and cognitive decline. It has been suggested to be of potential benefit in Chronic Fatigue Syndrome (CFS) by increasing blood flow [30]. St. John's Wort (*Hypericum perforatum*) contains hyperforins, valid for treatment of mild depression [31, 32]. Valerian root, *Valeriana officinalis*, has been used as a drug for more than 1,000 years. It is popular around the world as a sedative and mild hypnotic [32]. It contains iridoid glycosides, terpenoids, valerianic and isovaleric acid, and is used as an anxiolytic, for sleep improvement [30]. Kava kava, *Piper methysticum*, contains kavalactones and is used as an anxiolytic [30]. Some studies [32] suggest that kava may be effective for mild anxiety states.

Hypercholesterolemia

Plants sterols and stanols (phytosterols and phytostanols) are known to lower cholesterol levels in blood [33]. Some plants, such as mulberry (*Morus alba)* leaves, fruit, and bark; pomegranate (*Punica granatum*) fruit and peel; barley (*Hordeum vulgare*); date palm (*Phoenix dactylifera*); and garlic (*Allium sativum)* have been studied in this sense [34].

Soy-based supplements may play a valuable role in reducing cardiovascular risk. Soy (*Glycine max*) products containing isoflavones, cotyledon soy fibre, and soy phospholipids (Abaco and Abalon) significantly reduce the LDL:HDL ratio. Soluble dietary fibres such as psyllium and beta glucan from oat bran had a variable effect on LDL-cholesterol levels in the studies analyzed. Plant sterol esters, when consumed in margarines, lower the LDL: HDL ratio [35].

Garlic, *Allium sativum*, contains allicin. It is used to lower high cholesterol, hypertension, and in respiratory infections [31]. Another author [11] cites plant remedies to treat hypercholesterolemia such as sulphur derivates from *Allium sativum* [36, 37], mucilage found in algae such as glucomannan, ispagula, focus and zaragatona [38] [39]; and phytosterols such as sitosterol, campesterol and β−sitosterol. Sulphur active principles mainly in the form of cysteine derivatives in onion and garlic are responsible for biological activities such as anti-diabetic, antibiotic, hypocholesterolaemic and fibrinolytic [25, 40].

Malaria, the Challenge of Epidemics

Quinine, the alkaloid drug extracted from *Cinchona* spp bark, was the first drug used widely against malaria chemoprophylaxis from the 19th century. Quinine has been the main treatment for severe malaria since the introduction of *Cinchona* bark to European medicine in the 1630s. There is some evidence of drug resistance and decline in the efficacy of the treatment that is nowadays combined with other treatments, such as artemisinin, from *Artemisia annua*, and its semisynthetic derivates.

Artemisinin, or Chinese wormwood, is the basis of the most effective anti-malarials. This drug was "rediscovered" by 2015 Nobel Prize Tu Youyou, who in the early 1970s analyzed the efficacy of some hundreds of plant extracts from Chinese traditional medicine used against malaria, artemisinin being the most effective one. Artemisinin is effective also to combat other diseases and has demonstrated significant potential for the treatment of cancer and schistosomiasis. Nowadays, there are some reports on how the malaria parasite has become resistant to treatment with artemisinin [41]. Recently, some authors [42] report on the novel antiplasmodial activity of 4 previously described natural products [43]. Among the main ones available, a semisynthetic drug from artemisin, artesunate, is the treatment of choice. Artemether should only be used if artesunate is unavailable. Quinine dihydrochloride should be given only when artesunate and artemether are unavailable [44]. Research is done into natural products, also coming from fungus and microorganisms

Menopause and Momen's Health

Natural substances used to treat menopause symptoms are, for example, isoflavones such as genisteine and daizdeine [11].

Evening primrose (*Oenothera biennis,* Onagraceae) is reported to relieve the discomforts of PMS, menopause, menstruation, endometriosis and fibrocystic breasts by interfering with the production of inflammatory prostaglandins released during menstruation [45] [46]. Grape seed (*Vitis vinifera,* Vitaceae) proanthocyanidins and flavonoids are reported to be effective in improving the physical and psychological symptoms of menopause increasing muscle mass and reducing blood pressure in middle-aged women [46, 47]. Valerian root, *Valeriana officinalis* (Valerianaceae), has been used for centuries for its tranquilizing, menstruation, and sedative effects. Currently, valerian is prescribed for reducing pain, cyclic cramps, anxiety, and stress associated with dysmenorrhea [46, 48]. Chasteberry, *Vitex agnus-castus* (Verbenaceae), is one of the most popular and effective herbs used for the prevention and treatment of pre-menstrual syndrome (PMS) and hyperprolactinemia because of its hormone-like effects [46, 48]. Soy (*Glycine max*, Fabaceae) is used for the reduction of menopausal symptoms. Soy extracts contain isoflavones belonging to the group of phytoestrogens (daidzein, glycitein, genistein and their glycosides), which play an important role in reducing climacteric symptoms in menopausal and postmenopausal women [46]. Black cohosh root, *Cimicifuga racemosa*, contains triterpene glycosides that relief premenstrual symptoms. It has been used for the treatment of menopausal disorders in Germany for over 50 years, and is presently available in USA as a dietary supplement [10, 46].

Natural Antibiotics

The interest in the development of plant antimicrobials started in 1990s, when the efficacy of antibiotics started to decrease, they presented no desirable side effects and toxicity, and many microbials developed resistance against them [49].

Many species have been described as antimicrobial and natural antibiotic drugs, such as tea tree (*Melaleuca alternifolia*), lemon balm (*Melissa officinalis*), clove (*Syzygium aromaticum*), rosemary (*Rosmarinus officinalis*) and oregano (*Origanum vulgare*) [9]. Garlic (*Allium sativum*) is widely used for its antimicrobial properties. Garlic extracts are known to inhibit the growth of fungi, protozoa, viruses, and numerous bacteria, such as *Salmonella* spp., *Staphylococcus aureus*, and *Escherichia coli*. Garlic's antimicrobial *activity* is due to its bioactive compound allicin, considered a substitute or co-adjuvant for commercial antibiotics [50].

Different families of secondary metabolites exhibit antimicrobial activity [49]. Phenolics such as cinnamic and caffeic acids, catechol and pyrogallol are examples of them. The more hydroxylated they are, the stronger the antimicrobial activity they exert. Quinones and their main group, anthraquinones, act at bacterial cell wall and membrane. Stilbenes occur as glycosides, especially in the Orchidaceae, Dioscoraceae and Combretaceae families. Combrestatin 5 is an example of antimicrobial from *Combretum woodii*. Lipophilic flavonoids disrupt bacterial membranes. Their low toxicity makes them ideal antimicrobial medicines. Tannins are also known to inhibit bacterial and fungal activity. Coumarins and their relatives, hydroxycinnamic acids and phytoalexins, have antimicrobial activity, as in *Gallium odoratum* extracts. Terpenoids are also antimicrobial agents, as well as alkaloids, such as Solanargine from *Solanum khasia* or drymaritin from *Drymaria diandra*, which have antiviral activity against HIV.

These natural antimicrobials have been studied for their synergistic effect together with common antibiotics. For instance, 1,8 cineol, linalool, alpha-terpineol and terpinen-4-ol from *Melaleuca* spp. essential oils; alpha-Mangostin (*Garcinia mangostana*); Baicalin from *Scutellaria baicalensis;* Berberine from *Berberis* plant; Blueberry from *Vaccinium myrtillus*; Carnosic acid and Carnosol from *Rosmarinus officinalis* extract; Catechins, Epigallocatechin-gallate, and Butylated hydroxyanisole, (BHA) from green tea extract (*Camellia sinensis*); Corilagin from *Arctostaphylos uva-ursi;* Craneberry extract from *Vaccinium macrocarpon;* diterpenes from *Lycopus europaeus;* Erybraedin A or eryzerin C from the roots of *Erythrina zeyheri*; essential oils of *Leucodendron* spp. and *Ocimum gratissimum;* ferulenol from *Ferula communis*; Grape seed extract (*Vitis vinifera*); Isoflavone Bidwillon B from *Erythrina variegata*; isopimaric acid from *Pinus nigra;* Khat (*Catha edulis*) aqueous crude extracts; Myricetin; Novoimaninf from *Hypericum perforatum;* Oregano (*Origanum vulgare*) extract; plumbagin (from *Plumbago zeylanica*); Pomegranate (*Punica granatum*) extract; pyridine from *Jatropha elliptica*; Rugosin B and Tellimagrandin I from red rose tree, *Hibiscus rosa-sinensis*; flavanone G, from *Sophora japonica*; Totarol, from *Podocarpus totara* [51].

ESSENTIAL OILS, AROMATHERAPY AND HEALTH CARE

In the pharmaceutical industry, mint is used in medicinal tablets and potions, while capsicum (seeds) is used for thermo-plasters or for creams or gels against muscular pain. Children's medicine in particular is often designed to taste pleasant, so changes in taste may be an opportunity for further growth in the use of spices and herbs in medical applications. Restaurants, hospitals and schools are some of the other end-users of spices, although the quantities consumed by these sectors are small [52].

Aromatherapy is the use of essentials oils for therapeutic purposes. The absorption can be topical, oral or by inhalation, the latter having the fastest effect.

Because the molecules bind to olfactory bulb receptors they have an immediate effect on the hypothalamus and amygdala. In fact, the liposoluble components are able to penetrate the blood stream by inhalation. In aromatherapy, essential oils are obtained from raw materials by distillation or pressure. These are administered by inhalation or as external rubs or topically and have several different beneficial effects for human beings [53].

Essential oils have been studied for their properties in surgical patients to treat nausea, pre-surgical anxiety, post-surgical nausea, pain and infections. Orange (*Citrus sinensis* fruit oil), neroli (*Citrus sinensis* flower oil), mandarine orange (*Citrus paradisii* fruit oil), and lemon (*Citrus limon*) have been used for anxiety, lavender (*Lavandula officinalis*) for pain, and tea tree oil (*Melaleuca alternifolia*) for pre- and post-operative infections, cold, fever, flu, chickenpox and herpes [53] [54].

Essential oils have been studied for different microbials such as *Pseudomonas, E. coli, Streptococcus, Acinetobacter, Klebsiella, Chlostridium* [55].

Other plants used for aromatherapy are clary sage (*Salvia sclarea*), for women's health problems such as womb, uterus and menstrual disorders, eucalyptus (*Eucalyptus globulus*) for respiratory problems but also for skin problems, geranium (*Pelargonium graveolens*) is a sedative and for skin treatment, peppermint (*Mentha piperita*) is anti-inflammatory, analgesic, digestive and antimicrobial, roman chamomile (*Anthemis nobilis*) against hay fever, inflammation, muscle spasms, menstrual disorders, insomnia, ulcers, wounds, gastrointestinal disorders, rheumatic pain, and hemorrhoids, rosemary (*Rosmarinus officinalis*) digestive, cardiovascular, antirheumatic, skin tonic and hair growth improving, and ylang–ylang (*Cananga odorata*) against anxiety, depression, anxiety, hypertension, stress and it is also an aphrodisiac [53].

Table **1** shows essential oils studied for different affections and diseases [55].

Table 1. Essential oils studied on different affections [55].

Affection	Essential Oil
Addictions	Nicotine: *Angelica archangelica, Lavandula angustifolia, Piper nigrum*
	Alcohol: *Citrus bergamia, Chamaemelum nobile, Lavandula angustifolia*
	Opiates: *Mentha piperita*
Insomnia	*Citrus sinensis, Lavandula angustifolia, Rosa damascena*
Nausea and vomiting	*Carum carvi, Citrus sinensis, Elettaria cardamommon, Foeniculum vulgare Mentha piperita, Pimpinella anisum, Syzygium aromaticum, Zingiber officinale.*

(Table 1) contd.....

Affection	Essential Oil
Pain	*Aloysia triphylla, Boswellia carteri, Cananga odorata, Commiphora molmol, Cymbopogon citratus, Juniperus communis, Lavandula latifolia, L. angustifolia, Mentha piperita, Origanum majorana, Pepper nigrum, Rosa damascena, Syzygium aromaticum*
Respiratory	*Cinammomum camphora* CT cineole, *Eucalyptus globulus, E. radiata, E. smithii, Lavandula latifolia, Melaleuca alternifolia, Mentha piperita, Myrtus communis, Pinus sylvestris, Rosmarinus officinalis* CT cineole, *Thymus mastichina*
Stress	*Boswellia carteri, Citrus aurantium var. amara, Citrus bergamia, Citrus sinensis, Lavandula angustifolia, L x intermedia, Rosa damascena*
Skin	Acne: *Citrus sinensis, Ocimum basilicum*
	Athlete's foot: *Melaleuca alternifolia*
	Herpes: *Eucalyptus globulus, Melaleuca alternifolia, Thymus vulgaris*
	Wrinkles: *Boswellia serrata, Camellia japonica, Commiphora wightii, Rosmarinus officinalis, Zingiber officinale,*
Women's health	Dysmenorrhea: *Chamaemelum nobile, Lavandula angustifolia, Mentha piperita, Rosmarinus officinalis, Salvia sclarea*
	Menopause: *Jasminum officinale, Lavandula angustifolia, Pelargonium graveolens, Rosa damascena*

PLANTS FOR VETERINARY MEDICINE

Plant secondary metabolites have been used for ethnoveterinary medicine since ancient times. Nowadays, there is a growing interest in reducing chemicals in agriculture, due to their toxicity to animals, humans and the environment. Secondary metabolites such as saponins and tannins are being studied for ruminants, to enhance milk and meat quality, for nematodes control or methane production [56].

This industry uses the essential oil of *Chenopodium ambrosoides* prized for its Ascaridole content, for worming or helminthiasis. It also requires limonene and menthol as insecticides. The essential oil of thyme contains thymol which protects the bee barroa. Studies have been done on thyme oil as an alternative to antibiotics as growth promoters in birds (quail, chickens, turkeys). Lemon balm (*Melissa officinalis*), Garlic (*Allium sativum*) and Tea tree (*Melaleuca alternifolia*) rosemary and oregano have been described as broad spectrum antimicrobials. There is an opportunity for natural ingredients especially in products directed at domestic pets and horses, food supplements and natural antibiotics described in [9].

Oregano (*Origanum majoricum*) essential oil, for example, may be applied in animal feed, in the treatment of coccidiosis of domestic animals and candidiasis. It can be applied as a larvicide, repellent, insecticide and acaricide. It is used in aquaculture to treat fish diseases caused by bacteria and parasites or in the hatchery industry as a disinfectant for eggs or for disinfection of manure. The greatest potential of oregano essential oil is the possibility of its application in organic agriculture and organic animal husbandry [57].

According to Landato & Capasso [58], cattle, horses, sheep and goats represent the highest number of animal treated (70%) with medicinal plants, followed by poultry, (9,1%), dogs (5,3%) and rabbits (4,3%). Table **2** shows several examples of medicinal plants used to treat different veterinary affections and diseases [58].

Table 2. Plants of veterinary use [58].

Affection	Plants Used
Cardiotonic	*Coriandrum sativum, Crataegus monogyna*
Digestive	*Acacia catechu, Aloe* spp., *Artemisia* spp., *Cynara scolimus, Gentiana lutea, Glycirrhiza glabra, Hammamelis virginiana, Humulus lupulus, Melissa officinalis, Mentha piperita, Plantago* spp., *Pneumus boldus, Taraxacum officinalis, Ulmus glabra, Zingiber officinale*
Endoparasites	*Gossypium* spp., *Lavandula officinalis, Mentha piperita, Onobrychis viciifolia, Salvia officinalis, Urtica dioica*
Eyes	*Euphrasia officinalis, Matricaria chamomilla, Lotus corniculatus*
Helminthiasis	In Europe: *Allium sativum, Coriandrum sativum, Eucalyptus globulus, Matricaria chamomilla, Nicotiana tabaccum, Sinapis alba, Ruta graveolens, Zingiber officinale*
	In Africa: Anogeissus leiocarpus, *Cissampelos mucronata, Croton macrostachys, Khaya senegalensis, Securineja virosa, Senecio lyratipartus*
Hepatic and Immune system	*Echinacea* spp., *Malva sylvestris, Sylibum marianum*
Parasites	*Alllium sativum, Artemisia* spp., *Carica papaya, Cichorium intybus, Coriandrum sativum, Cucurbita pepo, Dorycnium* spp., *Eucalyptus* spp.
Reproductive system	*Capsella bursa-pastoris, Chrysantemum partemium, Hedera helix, Mentha pulegium*
Respiratory	*Borago officinalis, Eucalyptus globulus, Malva sylvestris, Sambucus nigra, Thymus vulgaris, Tilia cordata*
Skin	*Aloe* spp., *Anethum graveolens, Calendula officinalis, Carum carvi, Coriandrum sativum, Chrysantemum cinerariaefolium, Laurus nobilis, Mentha piperita, Matricaria chamomilla, Quassia amara*
Stress	*Humulus lupulos, Lavandula officinalis, Melissa officinalis, Passiflora incarnate, Valeriana officinalis*
Urinary	*Arctostaphyllos uva-ursi*
Wounds	*Echinacea* spp., *Calendula officinalis, Filipendula ulmaria, Hypericum perforatum, Lavandula officinalis, Plantago* spp., *Thymus vulgaris, Urtica dioica*

CONFLICT OF INTEREST

The authors declare no conflict of interest, financial or otherwise.

ACKNOWLEDGEMENTS

Declared none.

REFERENCES

[1] Houghton PJ. Old yet new—pharmaceuticals from plants. J Chem Educ 2001; 78: 175-84. [http://dx.doi.org/10.1021/ed078p175]

[2] Balandrin MF, Klocke JA, Wurtele ES, Bollinger WH. Natural plant chemicals: sources of industrial and medicinal materials. Science 1985; 228(4704): 1154-60. [http://dx.doi.org/10.1126/science.3890182] [PMID: 3890182]

[3] O'Hara M, Kiefer D, Farrell K, Kemper K. A review of 12 commonly used medicinal herbs. Arch Fam Med 1998; 7(6): 523-36. [http://dx.doi.org/10.1001/archfami.7.6.523] [PMID: 9821826]

[4] Attele AS, Wu JA, Yuan CS. Ginseng pharmacology: multiple constituents and multiple actions. Biochem Pharmacol 1999; 58(11): 1685-93. [http://dx.doi.org/10.1016/S0006-2952(99)00212-9] [PMID: 10571242]

[5] WHO. WHO Guidelines on Good Agricultural and Collection Practices (GACP) for Medicinal Plants. Geneva: World Health Organization 2003.

[6] Mishra BB, Tiwari VK. Natural products: an evolving role in future drug discovery. Eur J Med Chem 2011; 46(10): 4769-807. [http://dx.doi.org/10.1016/j.ejmech.2011.07.057] [PMID: 21889825]

[7] Ben Amar M. Cannabinoids in medicine: A review of their therapeutic potential. J Ethnopharmacol 2006; 105(1-2): 1-25. [http://dx.doi.org/10.1016/j.jep.2006.02.001] [PMID: 16540272]

[8] Hazekamp A, Grotenhermen F. Review on clinical studies with cannabis and cannabinoids 2005-2009. Cannabinoids 2010; 5: 1-21.

[9] CBI. CBI Market Survey. Trends Natural Ingredients for Health Products in Europe 2015.

[10] Lubbe A, Verpoorte R. Cultivation of medicinal and aromatic plants for specialty industrial materials. Ind Crops Prod 2011; 34(1): 785-801. [http://dx.doi.org/10.1016/j.indcrop.2011.01.019]

[11] Navarro Moll MC. Los productos naturales en la innovación farmacológica. Revista de Fitoterapia 2005; 5(1): 41-3.

[12] Balunas MJ, Kinghorn AD. Drug discovery from medicinal plants. Life Sci 2005; 78(5): 431-41. [http://dx.doi.org/10.1016/j.lfs.2005.09.012] [PMID: 16198377]

[13] van Der Heijden R, Jacobs DI, Snoeijer W, Hallard D, Verpoorte R. The *Catharanthus* alkaloids: pharmacognosy and biotechnology. Curr Med Chem 2004; 11(5): 607-28. [http://dx.doi.org/10.2174/0929867043455846] [PMID: 15032608]

[14] Xu M, Floyd HS, Greth SM, *et al.* Perillyl alcohol-mediated inhibition of lung cancer cell line proliferation: potential mechanisms for its chemotherapeutic effects. Toxicol Appl Pharmacol 2004; 195(2): 232-46. [http://dx.doi.org/10.1016/j.taap.2003.11.013] [PMID: 14998688]

[15] Lin SB, Li CH, Lee SS, Kan LS. Triterpene-enriched extracts from Ganoderma lucidum inhibit growth

of hepatoma cells *via* suppressing protein kinase C, activating mitogen-activated protein kinases and G2-phase cell cycle arrest. Life Sci 2003; 72(21): 2381-90. [http://dx.doi.org/10.1016/S0024-3205(03)00124-3] [PMID: 12639703]

[16] Muñoz SE, Piegari M, Guzmán CA, Eynard AR. Differential effects of dietary Oenothera, Zizyphus mistol, and corn oils, and essential fatty acid deficiency on the progression of a murine mammary gland adenocarcinoma. Nutrition 1999; 15(3): 208-12. [http://dx.doi.org/10.1016/S0899-9007(98)00181-6] [PMID: 10198915]

[17] Lau L, Supko JG, Blaney S, *et al.* A phase I and pharmacokinetic study of ecteinascidin-743 (Yondelis) in children with refractory solid tumors. A Children's Oncology Group study. Clin Cancer Res 2005; 11(2 Pt 1): 672-7. [PMID: 15701855]

[18] Bailly C. Homocamptothecins: potent topoisomerase I inhibitors and promising anticancer drugs. Crit Rev Oncol Hematol 2003; 45(1): 91-108. [http://dx.doi.org/10.1016/S1040-8428(02)00090-2] [PMID: 12482574]

[19] Lawrence NJ, Rennison D, McGown AT, Hadfield JA. The total synthesis of an aurone isolated from *Uvaria hamiltonii*: aurones and flavones as anticancer agents. Bioorg Med Chem Lett 2003; 13(21): 3759-63. [http://dx.doi.org/10.1016/j.bmcl.2003.07.003] [PMID: 14552774]

[20] Tozer GM, Prise VE, Wilson J, *et al.* Combretastatin A-4 phosphate as a tumor vascular-targeting agent: early effects in tumors and normal tissues. Cancer Res 1999; 59(7): 1626-34. [PMID: 10197639]

[21] Kaplan SA. Updated meta-analysis of clinical trials of *Serenoarepens* extract in the treatment of symptomatic benign prostatic hyperplasia. J Urol 2005; 173: 516. [http://dx.doi.org/10.1016/S0022-5347(05)60530-7]

[22] Dahanukar RA, Kulkarni NN, Rege SA. Pharmacology of Medicinal Plants and Natural Products. Indian J Pharmacol 2000; 32: 81-118.

[23] Ji HF, Li XJ, Zhang HY. Natural products and drug discovery. Can thousands of years of ancient medical knowledge lead us to new and powerful drug combinations in the fight against cancer and dementia? EMBO Reports 103 2009; 194-200.

[24] Agrawal P, Rai V, Singh RB. Randomized placebo-controlled, single blind trial of holy basil leaves in patients with noninsulin-dependent diabetes mellitus. Int J Clin Pharmacol Ther 1996; 34(9): 406-9. [PMID: 8880292]

[25] Nabavi SF, Thiagarajan R, Rastrelli L, *et al.* Curcumin: a natural product for diabetes and its complications. Curr Top Med Chem 2015; 15(23): 2445-55. [http://dx.doi.org/10.2174/1568026615666150619142519] [PMID: 26088351]

[26] Hardie DG. AMPK: a target for drugs and natural products with effects on both diabetes and cancer. Diabetes 2013; 62(7): 2164-72. [http://dx.doi.org/10.2337/db13-0368] [PMID: 23801715]

[27] Kelly GS. *Rhodiola rosea*: a possible plant adaptogen. Altern Med Rev 2001; 6(3): 293-302. [PMID: 11410073]

[28] Panossian A, Wikman G, Sarris J. Rosenroot (*Rhodiola rosea*): traditional use, chemical composition, pharmacology and clinical efficacy. Phytomedicine 2010; 17(7): 481-93. [http://dx.doi.org/10.1016/j.phymed.2010.02.002] [PMID: 20378318]

[29] Baschetti R. Chronic fatigue syndrome and liquorice. N Z Med J 1995; 108(998): 156-7. [PMID: 7761058]

[30] Wesnes KA, Faleni RA, Hefting NR, *et al.* The cognitive, subjective, and physical effects of a ginkgo biloba/panax ginseng combination in healthy volunteers with neurasthenic complaints. Psychopharmacol Bull 1997; 33(4): 677-83.

[PMID: 9493479]

[31] Bilia AR, Gallori S, Vincieri FF. St. John's wort and depression: efficacy, safety and tolerability-an update. Life Sci 2002; 70(26): 3077-96.
[http://dx.doi.org/10.1016/S0024-3205(02)01566-7] [PMID: 12008092]

[32] Mischoulon D. The herbal anxiolytics Kava and Valerian for anxiety and insomnia. Psychiatr Ann 2002; 32(1): 55-60.
[http://dx.doi.org/10.3928/0048-5713-20020101-09]

[33] Plat J, Mensink RP. Plant stanol and sterol esters in the control of blood cholesterol levels: mechanism and safety aspects. Am J Cardiol 2005; 96(1A): 15D-22D.
[http://dx.doi.org/10.1016/j.amjcard.2005.03.015] [PMID: 15992511]

[34] El-Sayyad HI. Cholesterol overload impairing cerebellar function: the promise of natural products. Nutrition 2015; 31(5): 621-30.
[http://dx.doi.org/10.1016/j.nut.2014.10.017] [PMID: 25837204]

[35] Hermansen K, Dinesen B, Hoie LH, Morgenstern E, Gruenwald J. Effects of soy and other natural products on LDL:HDL ratio and other lipid parameters: a literature review. Adv Ther 2003; 20(1): 50-78.
[http://dx.doi.org/10.1007/BF02850119] [PMID: 12772818]

[36] Lau BH. Suppression of LDL oxidation by garlic. J Nutr 2001; 131(3s): 985S-8S.
[PMID: 11238802]

[37] Banerjee SK, Maulik SK. Effect of garlic on cardiovascular disorders: a review http://www.nutritionj.com/content1/1/4
[http://dx.doi.org/10.1186/1475-2891-1-4]

[38] Mc Mahon M. Tratment guidelines for hipercolesterolemia: time toconsider solubre fibre. Int J Clin Pract 1999; 1999(53): 369-72.

[39] Davidson MH, Maki KC, Kong JC, *et al.* Long-term effects of consuming foods containing psyllium seed husk on serum lipids in subjects with hypercholesterolemia. Am J Clin Nutr 1998; 67(3): 367-76.
[PMID: 9497178]

[40] Augusti KT. Therapeutic values of onion (*Allium cepa* L.) and garlic (*Allium sativum* L.). Indian J Exp Biol 1996; 34(7): 634-40.
[PMID: 8979497]

[41] Ortiz Melon JM. The Lasker Award, traditional Chinese Medicine and the discovery of Artemisin. Anales de la Real Academia Nacional de Farmacia 2011; 77(4): 13-6.

[42] Pérez-Moreno G, Cantizani J, Sánchez-Carrasco P, *et al.* Discovery of New Compounds Active against Plasmodium falciparum by High Throughput Screening of Microbial Natural Products. PLoS One 2016; 11(1): e0145812.
[http://dx.doi.org/10.1371/journal.pone.0145812] [PMID: 26735308]

[43] Turschner S, Efferth T. Drug resistance in *Plasmodium*: natural products in the fight against malaria. Mini Rev Med Chem 2009; 9(2): 206-2124.
[http://dx.doi.org/10.2174/138955709787316074] [PMID: 19200025]

[44] White NJ, Pukrittayakamee S, Hien TT, Faiz MA, Mokuolu OA, Dondorp AM. Malaria. Lancet 2014; 383(9918): 723-35.
[http://dx.doi.org/10.1016/S0140-6736(13)60024-0] [PMID: 23953767]

[45] Sekeroglu N. Evening primrose: Oenotherabiennis L In: Awaad AS, Singh VK, Govil JN, Eds. Recent Progress in Medicinal Plants. Houston: Studium Press LLC 2010; vol. 28: pp. 143-58.

[46] Masullo M, Montoro P, Mari A, Pizza C, Piacente S. Medicinal plants in the treatment of women's disorders: Analytical strategies to assure quality, safety and efficacy. J Pharm Biomed Anal 2015; 113: 189-211.

[http://dx.doi.org/10.1016/j.jpba.2015.03.020] [PMID: 25863356]

[47] Terauchi M, Horiguchi N, Kajiyama A, *et al.* Effects of grape seed proanthocyanidin extract on menopausal symptoms, body composition, and cardiovascular parameters in middle-aged women: a randomized, double-blind, placebo-controlled pilot study. Menopause 2014; 21(9): 990-6. [http://dx.doi.org/10.1097/GME.0000000000000200] [PMID: 24518152]

[48] Mirabi P, Dolatian M, Mojab F, Majd HA. Effects of valerian on the severity and systemic manifestations of dysmenorrhea. Int J Gynaecol Obstet 2011; 115(3): 285-8. [http://dx.doi.org/10.1016/j.ijgo.2011.06.022] [PMID: 21959068]

[49] Panda SK, Rath CC. Phytochemicals as Natural Antimicrobials: Prospects and Challenges.Bioactive Phytochemicals - Perspectives for Modern Medicine, 1. New Delhi: Daya Publishing House 2012; pp. 331-80.

[50] Fonseca MJ, Tavares F. Natural Antibiotics: A Hands-on Activity on Garlic's Antibiotic Properties. Am Biol Teach 2011; 73(6): 342-6. [http://dx.doi.org/10.1525/abt.2011.73.6.7]

[51] Hemaiswarya S, Kruthiventi AK, Doble M. Synergism between natural products and antibiotics against infectious diseases. Phytomedicine 2008; 15(8): 639-52. [http://dx.doi.org/10.1016/j.phymed.2008.06.008] [PMID: 18599280]

[52] CBI Market Survey. The spices and herbs market in the EU 2009.

[53] Naser BA, Al-Wabel A, Shams S, Ahamad A, Khan SA, Anwar F. Essential oils used in aromatherapy: A systemic review. Asian Pac J Trop Biomed 2015; 5(8): 601-11. [http://dx.doi.org/10.1016/j.apjtb.2015.05.007]

[54] Stea S, Beraudi A, De Pasquale D. Essential oils for complementary treatment of surgical patients: State of the art. Evidence - Based Complementary and Alternative Medicine 2014; 1-6.

[55] Buckle J. Aromatherapy and integrative healthcare Clinical Aromatherapy Essential Oils in Healthcare. 3rd Edition. chapter 5. Elsevier 2015; pp. 95-116.

[56] Rochfort S, Parker AJ, Dunshea FR. Plant bioactives for ruminant health and productivity. Phytochemistry 2008; 69(2): 299-322. [http://dx.doi.org/10.1016/j.phytochem.2007.08.017] [PMID: 17919666]

[57] Vučinić M, Nedeljković-Trailović J, Trailović S, Ivanović S, Milovanović M, Krnjaić D. Possibility for use essential oils in veterinary medicine and animal husbandry with special emphasis on oregano oil. Vet Glas 2012; 66(5-6): 407-16. [http://dx.doi.org/10.2298/VETGL1206407V]

[58] Landato M, Capasso R. Useful plants for animal therapy. OA Alternative Medicine 2013; 1(1): 1.

CHAPTER 6

Other Uses of MAPs

M. Paz Arraiza[1,*], Carlos Calderón-Guerrero[1], Silvia C. Guillén[2] and Miguel A. Sarmiento[3]

[1] *Department of Forestry and Environmental Engineering and Management. Technical University of Madrid. Spain*

[2] *Ecology Department, Agronomy Faculty, National University of Tucuman. Argentina*

[3] *Faculty of Forestry, National University of Santiago del Estero, Argentina*

Abstract: Aromatic and medicinal plants are not only used in the pharmaceutical, cosmetic and food processing industries, as we have seen, but they have many other uses, increasing in number every day as new uses are researched and discovered. These other uses of MAPs are very important due to their growing economic importance and market possibilities. Natural dyes, pesticides, fungicides and insecticides, gums, resins, turpentines, allelopathic substances and tannins are some of them. Plant extracts have been used as insecticides since before ancient Rome, a practice that continues today with more than 2,000 species. There are different strategies and the compounds, such as allopathic substances, can be found in plants of different species. Resins, tannins, gums and balsams have been widely used until today and new products are still based on their natural properties.

Keywords: Allopathic substances, Balsams, Crop protection, Dyes, Fungicides, Gums, Insect control, Insecticides, Pesticides, Resins, Tanning, Tannins, Turpentines.

INTRODUCTION

Aromatic and medicinal plants are not only used in the pharmaceutical, cosmetic and food processing industries, as we have seen, but they have many other uses, increasing in number every day as new uses are researched and discovered.

These other uses of MAPs are many, among which we have highlighted some, for their growing economic importance and market possibilities. Natural dyes, pesticides, fungicides and insecticides in agriculture, allopathic substances, resins, turpentines and tannins are some of them.

* **Corresponding author María Paz Arraiza:** Department of Forestry and Environmental Engineering and Management, Technical University of Madrid, Madrid, Spain; Tel/Fax: +34 91 336 64 08; Email: paz.arraiza@upm.es

M. Paz Arraiza, A. González-Coloma, J. Burillo and C. Calderón-Guerrero (Eds.)

NATURAL DYES

Compounds extracted from plants have been used as natural colorants or dyes for centuries. Since the industrial era, these compounds have been replaced by synthetic ones [1], much more effective, resistant and long-lasting, but often, also toxic and irritating. In recent times, there is a trend to return to natural sustainable products, in this case natural dyes, and extensive research is being done in this field [2].

Textile processing is one of the most pollutant industrial processes, needing about 250 tons of water per ton of textile dyed and generating the same amount of effluent [3]. According to this, there are two ways of minimizing this environmental impact, one is through the treatment and depuration of effluents, and the other is to look for environmentally friendly dyes [4].

Natural dyes are not only used to give colour to natural fibres (wool, cotton, linen and hemp) but also for paints and varnishes, cosmetics, food, buildings and in painting restoration [5].

Active ingredient families for these types of products are alkaloids, anthocyanins, anthraquinones, betains, carotenoids, flavonoids, naphthoquinones, polyphenols and tannins.

Some examples of these plants are indigo, woad, weld, madder, safflower, logwood and achiote, among many others. Indigo (*Indigofera tinctoria*) has been one of the most important dyestuffs [6], nowadays used to dye denim. Woad (*Isatis tinctoria*) contains a small amount of the same compounds as are found in indigo. Weld (*Reseda luteola*) gives a yellow colour. Changing the mordant of madder (*Rubia tinctoria*) can give red, pink, lilac, orange, black, and brown coloured pigments. Safflower (*Carthamus tinctorius*) makes a red or yellow dye. Logwood (*Haemotoxylon campechianum*) dyes things black at first. Achiote or annatto (*Bixa orellana*) was used by the Aztecs and Mayas. This substance is also now used to dye margarine and similar products. Table **1** shows a compilation of dye plants used worldwide.

Table 1. Some plants used for dyeing (source [1] modified).

Plantname	Scientificname	Family	Chemicals	Colours
Annatto	*Bixa orellana*	Bixaceae	Carotenoids	Yellow, orange
Ashtree	*Fraxinus excelsior*	Fagaceae	Flavonoids	Beige, black
Barberry	*Berberis vulgaris*	Berberidaceae	Alkaloids	Yellow, brown
Dyer's knot	*Polygonum tinctorium*	Polygonaceae	Alkaloids	Indigo blue

(Table 1) contd.....

Plantname	Scientificname	Family	Chemicals	Colours
Goldenrod	*Solidago* spp.	Asteraceae	Flavonoids	Yellow, olive
Grapes	*Vitis vinifera*	Vitaceae	Anthocyanins	Red, blue
Henna	*Lawsonia inermis*	Lythraceae	Naphthoquinones	Red
Hollyhock	*Alcea rosea*	Malvaceae	Anthocyanins	Brown, green
Indigo	*Indigofera tinctoria*	Fabaceae	Indigotin	Blue
Logwood	*Haemotoxylon campechianum*	Caesalpinaceae	Tannins	Black
Madre	*Rubia tinctorum*	Rubiaceae	Anthraquinones	Red, brown
Marigold	*Tagetes patula*	Asteraceae	Flavonoids	Yellow
Páprika	*Capsicum annuum*	Solanaceae	Carotenoids	Orange, red
Privet	*Ligustrum vulgare*	Oleaceae	Anthocyanins	Blue, green
Red beet	*Beta vulgaris*	Amaranthaceae	Betanins	Pink, red
Safflower	*Carthamus tinctorius*	Asteraceae	Flavonoids	Yellow, red
Stickyalder	*Betula alnus*	Betulaceae	Tannins	Beige, black
Turmeric	*Curcuma longa*	Zingiberaceae	Polyphenols	Bright yellow
Walnut	*Juglans regia*	Juglandaceae	Naphthoquinones	Brown
Weld	*Reseda luteola*	Resedaceae	Flavonoids	Yellow
Woad	*Isatis tinctoria*	Brassicaceae	Alkaloids	Indigo blue

PLANTS USED FOR CROP PROTECTION

Plant extracts have been used as insecticides since before ancient Rome, a practice that continues today with more than 2,000 species with such properties. These plants contain substances such as pyrethrins, rotenone and alkaloids which have been used as insecticides. Their use is particularly prevalent in developing countries, where local plants are cheaper than chemical insecticides. Also, chemical pesticides such as DDT have led to very serious environmental problems. Nowadays, the awareness of the importance of the use of environmental and health friendly substances has led to research in the field of natural pesticides. These plants are employed directly or as precursors for the production of pesticides [7, 8].

Plants Used as Antifungal Agents

Plants are a source of antifungal compounds. Some plants with antifungal effect against *Aspergillus* spp. are *Acacia nilotica, Achras zapota, Datura stramonium, Eucalyptus globulus, Lawsonia inermis* and *Punica granatum* among others [9]. Other families and species are being studied as a source of phytoalexins, such as the Leguminosae and Rosaceae families, and rice, *Oryza sativa* [10]. Extracts of

Allium spp., *Capsicum* spp., palmarrosa (*Cymbopogon martinii*), red thyme (*Thymus zygis*), cinnamon leaf (*Cinnamomum zeylanicum*) and clove (*Syzygium aromaticum*) have antifungal activity against *Botrytis cinerea*. The compounds that exert higher antifungal activity are D-limonene, cineole, β myrcene, α and β pinene and camphor [11].

Essential oils rich in carvacrol (*Origanum majoricum; O. dictamnus; O. majorana*), thymol (*Thymus vulgaris*), linalool and linalyl acetate (*Lavandula* spp.), eucalyptol (*Rosmarinus officinalis, Salvia officinalis*) and cis-menthone and pulegone (*Mentha pulegium*) have been tested for their effectiveness against *Botrytis cinerea, Fusarium* spp. and *Clavibacter michiganensis*, being the essential oils rich in carvacrol and thymol the most active against these fungi [12].

Jojoba (*Simmondsia californica*), rosemary (*Rosmarinus officinalis*) and thyme (*Thymus vulgaris*) essential oils have antifungal activity. A mixture of 31% *Gossypium hirsutum* and 23% *Allium sativum* oils is also employed against fungi in plant crops [8].

Plants Used for Weed Control

Plant extracts, natural oils and essential oils are used for weed control especially in organic agriculture.

Essential oils used for weed control in crop protection are chenopodium (*Chenopodium album*), containing p-cymene, γ−terpinene and thymol [8]; clove (*Syzyngium aromaticum*) oil containing eugenol as the main component; cinnamon oil (*Cinnamomum zeylanicum*); citronella oil (*Cymbopogon* spp.) containing citronellal and geraniol; citrus oil (*Citrus* spp.); *Eucalyptus citriodora* rich in citronellal; garlic oil (*Allium sativum*); lemongrass oil (*Cymbopogon citratus*) containing mainly citral; pine essential oil (*Pinus* spp.); oregano (*Origanum* spp.). Some compounds isolated from essential oils used for weed control are D-limonene, found in many species, which removes the wax cuticle of plants causing dehydration, and phenethyl propionate, from *Mentha piperita* [8].

Vegetable oils used for weed control are castor oil (*Ricinus communis*), coconut oil (*Cocus nucifera*), niger oil (*Guizotia abyssinica*), mustard oil (*Brassica juncea*) and sunflower oil (*Helianthus annuus*), also applied for organic crop production [8].

Allelopathic Substances

Allelopathic substances are produced by plants to selectively inhibit the growth of other plants or microorganisms in the soil surrounding them. These phytotoxic compounds have a roll in the allelopathic interactions among plants and they act as natural herbicides, microbial inhibitors and seed germination inhibitors as well. Although many allelochemical substances are defence substances, some others act directly against weeds, by competence and density control. These types of substances include volatile terpenoids, phenylpropanoids, quinones, coumarins, flavonoids, tannins, phenolics and glycosides. Momilactone B in *Oryza* spp., sorgoleone and benzoquinone from *Sorghum bicolor*, and benzoxazionids in the Poaceae family are some examples [8]. They are not commercially produced yet, but they can serve as a model for the new synthetic herbicides and antimicrobials.

Plants Used for Insect Control

Many substances derived from plants are currently being used to control insect pests, the most important ones being pyrethrins, extracted from *Chrysanthemum cinerariaefolium,* achenes, rotenone isolated from *Derris lanchocarpus* roots and azadirachtins isolated from seeds of the Indian neem tree (*Azadirachta indica*) [13] and some essential oils from aromatic plants [14].

Pyrethrins

Pyrethrins are the insecticidal ingredients occurring in the flowers of *Chrysanthemum cinerariaefolium*, which have been used widely for human and animal health protection by controlling insect pests such as cockroaches, houseflies, and mosquitoes since ancient times [8]. Natural pyrethrins consist of six compounds (pyrethrin I and II, jasmolin I and II, and cinerin I and II) [15]. They are the most economically important plant origin supply of natural insecticidal substances. They have been used since the 1800s. In the 1950s their use decreased due to the synthesis of similar chemical substances, such as allethrins, but they are still economically viable, and they represent a big area of this market. They are grown for commercial purposes in Kenya, Tanzania and Ecuador [16].

Rotenones

Rotenone is obtained from some plants of the Leguminosae species such *Derris* spp., *Lonchocarpus* spp. and *Tephrosia* spp. [8]. They are used as insecticides but also for fish population control, for they are toxic for fish, allowing their use for this purpose. Rotenone, one of the most famous botanical insecticides, possesses strong stomach poison, contact toxicity, and shows anti-feeding and growth inhibition activity. It has been proved that rotenone has such a special mode of action on insect pests that it could effectively control more than 800 kinds of insect pests infecting vegetables, fruit, rice, tea, tobacco, cotton, grass, and so on [17].

Alkaloids

Nicotine and its chemical relative nornicotine from *Nicotiana tabacum* (Solanaceae) are the most used alkaloids for insect control. They are used to combat mealybugs and aphids but are not accepted in certified organic production. Their use has decreased because of the high cost of production, nasty smell, extreme toxicity to mammals, and limited insecticidal activity. Other alkaloids are anabasine, from *Nicotiana glauca*, physostigmine, from calabar bean (*Physostigma venenosum*), which served as a model compound for the development of the carbamate insecticides, veratrine, a mixture of alkaloids (cevadine, veratridine, and sabacilline) and ryanodine [16].

Essential Oils

Some known essential oils produce substances used for insect and pest control. The terpenes limonene and linalool, found in citrus peels, are effective against fleas, aphids, mites, ants, flies, wasps and crickets. They are commercially sold as liquids, aerosols, shampoos, and dips for pets. Neem tree oil from *Azadirachta indica* contains insecticidal compounds, such as azadirachtin, which acts by preventing insect growth, and is thus useful to control insects in their growing stages. It is also used to control mildew. Ryania from *Ryania speciosa* is used against caterpillars and sold wholesale alone or mixed with rotenone and pyrethrum. Sabadila from *Schoenocaulon officinale* seeds is used against squash bugs, caterpillars, grasshoppers and fruit bugs.

Rosemary oil (*Rosmarinus officinalis*) contains 25% of 1,8-cineole, borneol, camphor and monoterpenoids, all constituents possessing insecticidal activity. It combats aphids, beetles, whiteflies, spider mites, thrips, and caterpillar larvae, among others. There are products available commercially for example mixed with peppermint oil (*Mentha piperita*) [8].

Thyme oil (*Thymus vulgaris*) with thymol and carvacrol and many monoterpenoids as main components is a broad-spectrum insect control in organic farming [8, 12]. Clove oil (*Syzygium aromaticum*) consists of 92% eugenol. It is a household and ornamental plants pest insecticide [8]. Capsaicin resin or powder from hot chilli peppers (*Capsicum frutescens*) contains approximately 3% capsaicin and is used as an insecticide [8].

Another very important use of essential oils is the organic control of the bee (*Apis mellifera*) mite Varroa (*Varroa destructor*). Thymol, isolated from thyme oil (*Thymus vulgaris*) is easy to use, safe for beekeepers, causes no honeybee toxicity, and causes low bee mortality [18].

Plant Growth Regulators

Brassinolides are compounds with a possible agricultural use. For example, a brassinolide extract from *Brassica napus* at low concentration promotes and accelerates plant growth. This compound and its synthetic relatives, brassinosteroids, may increase plant biomass through cell expansion and division. They increase biomass yield in crops with very small doses. They can be commercially produced by semi-synthesis the same way that the pharmaceutical steroid hormones are produced from steroid sapogenins.

GUMS, BALSAMS, RESINS AND TURPENTINES

Turpentine and Rosin

Turpentine is the resin obtained from conifers. Rosin is the solid residue after distilling turpentine and obtaining its volatile fraction. Turpentine and rosin have been used for naval stores (especially *Pinus* species, but also *Picea* and *Abies* species). Gum naval stores come from the tapping of living trees, sulphate naval stores come from the Kraft paper-making process, and wood naval stores come from pine stumps after the tree has been felled [19].

Gums

Gums are solids consisting of mixtures of carbohydrates which are either water-soluble or absorb water and swell up to form a gel or jelly when placed in water. They are insoluble in oils or organic solvents such as hydrocarbons, ether and alcohol. They are usually produced by exudation in response to injury to the plant.

Natural gums are employed in pharmaceutical uses, such as excipients, because they are economical, readily available, non-toxic, capable of chemical modifications, potentially biodegradable and biocompatible. There are exudates of gums and seed gums.

Exudate Gums

Gum Arabic from *Acacia senegal*, with food, pharmaceutical and technical uses. It is food additive E414, used in candy and soft drinks. Karaya gum is the dried exudate obtained from trees of the *Sterculia* species in Europe and it is the food additive E416 with strong adhesive properties used in foods and pharmaceuticals. Tragacanth from *Astragalus* species is used in foods and pharmaceuticals.

Seed Gums

Locust Bean Gum from the seeds from the carob tree *Ceratonia siliqua* contains galactomannan polysaccharides, used as a thickening and a gelling agent in food technology; Tara gum from seeds of *Caesalpinia spinosa* is used as a thickening agent and stabilizer in food applications.

Other examples of natural gums are Guar gum from the seed of the legume plant *Cyamopsis tetragonolobus*, used as a thickener in cosmetics, sauces, and as an agent in ice cream; Tamarind xyloglucan is obtained from the endosperm of the seed of the tamarind tree, *Tamarindus indica*, and used as stabilizer, thickener, gelling agent and binder in food and pharmaceutical industries; Okra gum, from the fruits of *Hibiscus esculentus*, has been studied as a binder in paracetamol tablet formulations; China rose or *Hibiscus rosa-sinensis*, has been studied in the development of sustained release tablets; Ishpagula or portions of the seeds of the plant *Plantago ovata*, is used as a binder, desintegrant and release retardant [20].

Resins and Balsams

Resins are solid or semi-solid materials of a complex mixture of terpenes, insoluble in water but soluble in certain organic solvents. Most resins of commercial interest are obtained as exudates by tapping. Balsams are resins characterized by a high content of benzoic or cinnamic acids and their esters. They have a characteristic "balsamic" odour.

Hard Resins

Copal comes from different tree species from Africa, Asia and South America (*Copaifera* spp. or *Hymenaea* spp.). Most copal of commercial interest comes from *Agathis* species of Southeast Asia. Copal resin has been used in the treatment of headaches, fever, burns and stomach ache. In dentistry, it is used as a binding media in dental products and in treatment of micro leakage in teeth, and recently, it has been evaluated as a matrix for drugs [19]. Mastic is the hard resin produced by tapping the tree (*Pistachia lentiscus,* Anacardiaceae). It is used in Arab countries for chewing, for it sweetens the breath and helps preserve the teeth and gums. It is also used as flavouring for alcoholic beverages, and its essential oil is used for fragrance and flavouring purposes. There are soft resins and balsams, and hard resins [21].

Soft Resins and Balsams

Benzoin is obtained from trees of the genus *Styrax* from Southeast Asia. There are two types, Siam benzoin from *S. tonkinensis* and Sumatra benzoin (also called gum Benjamin) from *S. benzoin*. Styrax, sometimes called storax, is a balsamic oleoresin obtained from trees of the genus *Liquidambar*. Peru and tolu balsams are obtained from the genus *Myroxylonbalsamum,* Leguminosae. Gum elemi or

Manila elemi is a fragrant oleoresin obtained from the trunk of *Canarium luzonicum*; all of them or the essential oil distilled from some of them are used in perfumery.

Latexes

Latex is a milky fluid, consisting of droplets of organic matter suspended or dispersed in an aqueous medium. The most well-known example is rubber latex, the coagulated latex obtained from the *Hevea brasiliensis* tree. Guttapercha is obtained from the plant *Palaquium gutta*. It has excellent insulating water-proof properties, making it very useful for insulating material for submarine and underground cables. It was also used for golf ball covers and other moulded products.

TANNING

Tanning is the process by which animal skin is converted into leather. This process needs the use of substances that help bind together the skin proteins in a flexible, stable and lasting way. Some plants produce these types of compounds, called tannins. There are two types of tannins, hydrolysable and condensed, the latter being the most important for the tanning process.

In Europe the most important tannin sources were chestnut (*Castanea sativa*), larch (*Larix decidua*), oak bark (*Quercus* spp.), galls (*Quercus infectoria*) and acorns (*Quercus macrolepis*), pomegranate (*Punica granatum*), spruce (*Picea abies*), sumac (*Rhus coriaria*) and wattle (*Acacia decurrens*).

In North America, the natives knew the use of tannins for leather and the colonists learned their uses, such as cainagre (*Rumex hymenosepalus*), chestnut (*Castanea dentata*), gambier (*Uncaria gambier*), hemlock (*Tsuga canadensis, T. heterophylla*), mangrove (*Rhizophora mangle*), oaks (*Quercus alba, Q. borealis, Q. montana, Q velutina*), and tanbark oak (*Lithocarpus densiflora*).

In South America, algarrobilla (*Caesalpinia brevifolia*), sumac (*Rhus glabra, R. typhina, R. copallina*) and quebracho (*Schinopsis lorentzii, S.balansae*). Some others are being grown, such as dividivi (*Caesalpinia coriaria*) and tara tree (*Caesalpinia spinosa*).

In Asia some plants used for tannins are avaram (*Cassia auriculata*) and myrobalan (*Terminalia chebula, T. bellerica*) from India, Chinese oak (*Rhus chinensis*), and gambier (*Uncaria gambier*) from Malaysia and Indonesia.

Other sources of tannins have been eucalyptus (*Eucalyptus occidentalis*) and Tanekaha (*Phyllocladus trichomanoides*) from Oceania and Palmeto root (*Sabal palmetto*). Table **2** shows several examples of plants used for tanning.

Table 2. Plants used for tanning

Scientific Name	Tree	Part Used	Origin
Acacia decurrens	Wattle	Bark	Europe
Caesalpinia brevifolia	Algarobilla	Fruit	Chile
Caesalpinia coriaria	DiviDivi	Fruit	West Indies, S. America
Caesalpinia spinosa	Tara	Fruit	Peru and N. Africa
Cassia auriculata	Avaram	Bark	India
Castanea sativa	Chestnut	Wood	Europe
Eucalyptus occidentalis	Eucalyptus	Bark	Australia
Larix decidua	European larch	Bark	Europe
Lithocarpus densiflora	Tanbark oak	Bark	North America
Phyllocladus trichomanoides	Tanekaha	Bark	New Zealand
Picea abies	Norway spruve	Bark	Europe
Punica granatum	Pomegranate	Fruit	Europe
Quercus alba	White oak	Bark	North America
Quercus borealis	Red oak	Bark	North America
Quercus infectoria	Aleppo oak	Galls	Europe
Quercus macrolepis	Turkish oak	Acorns	Turkey
Quercus montana	Chestnut oak	Bark	North America
Quercus robur	European oak	Bark	Europe
Quercus velutina	Black oak	Bark	North America
Rhizophora mangle	Mangrove	Bark	Tropics
Rhus chinensis	Chinese oak	Galls	China
Rhus coriaria	SicilianSumac	Leaves	Europe
Rhus glabra, R. typhina, R. copallina	American sumac	Leaves	South America
Rumexhy menosepalus	Cainagre	Root	US and Mexico
Sabal palmetto	Palmeto	Root	North America
Schinopsis lorentzii, S. balansae	Quebracho	Wood	South America
Terminalia chebula, T. bellerica	Myrobalan	Fruit	India
Tsuga canadensis	Eastern Hemlock	Bark	North America
Tsuga heterophylla	Western Hemlock	Bark	North America
Uncaria gambier	Gambier	Leaves	Malaysia Indonesia

CONFLICT OF INTEREST

The authors declare no conflict of interest, financial or otherwise.

ACKNOWLEDGEMENTS

Declared none.

REFERENCES

[1] Lubbe A, Verpoorte R. Cultivation of medicinal and aromatic plants for specialty industrial materials. Industrial Crops and Products 2011; 34(1): 785-801. [http://dx.doi.org/10.1016/j.indcrop.2011.01.019]

[2] Bechtold T, Turcanu A, Ganglberger E, Geissler S. Natural dyes in modern textile dyehouses—how to combine experiences of two centuries to meet the demands of the future? J Clean Prod 2003; 11: 499-509. [http://dx.doi.org/10.1016/S0959-6526(02)00077-X]

[3] Mirjalili M, Nazarpoor K, Karimi L. Eco-friendly dyeing of wool using natural dye from weld as co-partner with synthetic dye. J Clean Prod 2011; 19: 1045-51. [http://dx.doi.org/10.1016/j.jclepro.2011.02.001]

[4] Viana TC, Pagnan TS, Ayres E. Natural dyes in the design of textile: how to make them more competitive face to synthetic dyes. Journal of the International Colour Association 2015; 14: 14-27.

[5] CBI. EU Market Opportunities: The Use of Natural Dyes 2009.

[6] Gilbert KG, Cooke DT. Dyes from plants: past usage, present understanding and potential. Plant Growth Regul 2001; 34: 57-69. [http://dx.doi.org/10.1023/A:1013374618870]

[7] Copping LG, Duke SO. Natural products that have been used commercially as crop protection agents. Pest Manag Sci 2007; 63(6): 524-54. [http://dx.doi.org/10.1002/ps.1378] [PMID: 17487882]

[8] Dayan FE, Cantrell CL, Duke SO. Natural products in crop protection. Bioorg Med Chem 2009; 17(12): 4022-34. [http://dx.doi.org/10.1016/j.bmc.2009.01.046] [PMID: 19216080]

[9] Satish S, Mohana DC, Ranhavendra MP, Raveesha KA. Antifungal activity of some plant extracts against important seed borne pathogens of *Aspergillus* sp. Agric Technol Thail 2007; 3(1): 109-19.

[10] Grayer RJ, Kokubun T. Plant--fungal interactions: the search for phytoalexins and other antifungal compounds from higher plants. Phytochemistry 2001; 56(3): 253-63. [http://dx.doi.org/10.1016/S0031-9422(00)00450-7] [PMID: 11243452]

[11] Wilson CL, Solar JM, El Ghaouth A, Wisniewski ME. Rapid evaluation of plant extracts and essential oils for antifungal activity against Botrytis cinerea. Plant Dis 1997; 81(2): 204-10. [http://dx.doi.org/10.1094/PDIS.1997.81.2.204]

[12] Daferera DJ, Ziogas BN, Polissiou MG. The effectiveness of plant essential oils on the growth of *Botrytis cinerea, Fusarium* sp. and *Clavibacter michiganensis* subsp. *michiganensis.* Crop Prot 2003; 22(1): 39-44. [http://dx.doi.org/10.1016/S0261-2194(02)00095-9]

[13] Guleria S, Tiku AK. Botanicals in pest management: current status and future perpectives.Integrated Pest Management: Innovation-Development Process. Berlin: Springer 2009; Vol. 1: pp. 317-30. [http://dx.doi.org/10.1007/978-1-4020-8992-3_12]

[14] Isman MB. Botanical insecticides, deterrents, and repellents in modern agriculture and an increasingly regulated world. Annu Rev Entomol 2006; 51: 45-66. [http://dx.doi.org/10.1146/annurev.ento.51.110104.151146] [PMID: 16332203]

[15] Kaneko H. Pyrethroid Chemistry and Metabolism.Hayes' Handbook of Pesticide Toxicology. 3rd ed. New York: Academic Press 2010; pp. 1635-63. [http://dx.doi.org/10.1016/B978-0-12-374367-1.00076-8]

[16] Balandrin MF, Klocke JA, Wurtele ES, Bollinger WH. Natural plant chemicals: sources of industrial and medicinal materials. Science 1985; 228(4704): 1154-60. [http://dx.doi.org/10.1126/science.3890182] [PMID: 3890182]

[17] Huanong Biological Engineering Co. Ltd. Botanical insecticide rotenone. China Chemical Reporter 2005; 16(23)

[18] Jawad K, Asif A, Khalida R, Hamid A, Muhammad S. Thymol as control agent of mites (*Varroa destructor*) on honeybees (*Apis mellifera*). Pak J Agric Res 2013; 26(4)

[19] Ciesla WM. Products of resin processing. 2009.

[20] Avachat AM, Dash RR, Shrotriya SN. Recent investigations of plant based natural gums, mucilagues and resins in novel drug delivery systems. Ind J Pharm Edu Res 2011; 45(1): 86-99.

[21] Coppen JJ. Non-Wood Forest Products 6 Gums, resins, and latexes of plant origin. Rome: Food and Agriculture Organization of the United Nations 1995.

CHAPTER 7

Chemical Analysis and Quality Control of MAPs

M. Paz Arraiza[1,*], **Carlos Calderón-Guerrero**[1], **Silvia C. Guillén**[2] and **Miguel A. Sarmiento**[3]

[1] *Department of Forestry and Environmental Engineering and Management, Technical University of Madrid, Spain*

[2] *Ecology Department, Agronomy Faculty, National University of Tucuman, Argentina*

[3] *Faculty of Forestry, National University of Santiago del Estero, Argentina*

Abstract: The therapeutic activity of a plant medicine is determined by its quality, security and efficacy. These must be controlled in an exact, precise and repeatable process. Quality control is the set of technical operations employed to verify the fulfilment of the quality requirements established. The variability of the plant material depends on the identity of the plant, its origin (wild harvested or cultivated), the harvesting time, the soil and climatic conditions, the organ or part of the plant used, and the phenology and ontogeny of the plant. The plant or drug quality must be determined by identification of the plant material, botanically, physically and chemically.

Keywords: Aflatoxins, Ash, Chemical analysis, Chromatography, Distillation, Drug, Extraction, Foreign matter, Haemolitic activity, Pesticides, Quality control, Sampling, Swelling index, Tannins, Toxic metals.

INTRODUCTION

The therapeutic activity of a plant medicine is determined by its quality, security and efficacy. These must be controlled in an exact, precise and reproducible process. Quality control is the set of technical operations employed to verify the fulfilment of the quality requirements established. The variability of the plant material depends on the identity of the plant, its origin (wild harvested or cultivated), the harvesting time, the soil and climatic conditions, the organ or part of the plant used, and the phenology and ontogeny of the plant [1]. The plant or drug quality must be determined by identification of the plant material, botanically, physically and chemically.

* **Corresponding author María Paz Arraiza:** Department of Forestry and Environmental Engineering and Management, Technical University of Madrid, Madrid, Spain; Tel/Fax: +34 91 336 64 08; Email: paz.arraiza@upm.es

M. Paz Arraiza, A. González-Coloma, J. Burillo and C. Calderón-Guerrero (Eds.)

That, according to the World Health Organization (WHO) [2], is what standardization is, the confirmation of a plant material identity, determination of its quality, and the detection of the presence and nature of adulterants, by means of morphological, physical, chemical and biological methods.

These methods are the following [1]:

- Sampling and determination of foreign matter.
- Morphological, anatomical, and organoleptic tests.
- Macroscopic observation of the drug.
- Microscopic observation of the powdered drug.
- Thin layer chromatography.
- Qualitative and quantitative determination of active components. Extractable matter, foreign matter, ash, water and volatile matter, volatile oils, haemolytic activity, tannins, swelling index, radioactive contamination, aflatoxins, toxic metals, pesticide residues, micro-organisms.

SAMPLING

When analysing vegetable matter for use in pharmacology, quality control of samples becomes highly important, and the guidelines established by the Pharmacopoeias [3 - 6] or those of the World Health Organisation [7 - 9] must be followed, in a sample which is representative of the whole. Plant material is firstly powdered. According to the size of the mesh used to sieve it, powder can be coarse cut 4.00 mm, medium cut 2.80 mm and fine cut 2.00 mm [2]. Every sample should be inspected for organoleptic characteristics, presentation of the material, presence of foreign matter, presence of insects and presence of packaging material.

NATURE AND AMOUNT OF FOREIGN MATTER

Foreign matter is composed by either plant material different from that specified, organisms or their products (moulds, insects), or mineral products such as sand, glass, stones and dust. Normally, pharmacopoeias allow up to 2%. It is determined by sieving and macroscopic examination but microscopic examination should always be carried out in powdered material. An amount of sample is weighed and spread in a thin layer. The foreign matter is sorted into groups either by visual inspection, using a magnifying lens, or with a suitable sieve. Each group of foreign matter is weighed separately, and the percentage (W/W) of foreign matter is calculated as the sum of all of them.

MORPHOLOGICAL, ANATOMICAL, AND ORGANOLEPTIC TESTS

These tests serve to confirm the identity of the plant or drug, and to detect possible adulterations and falsifications.

Organoleptic Analysis

Organoleptic indicators include smell, colour, taste and texture.

Smell

It can be aromatic, garlicky, camphorated, nauseous, unpleasant, spicy, *etc.* Many plants and drugs have characteristic smells, *i.e.* mint, anise, cinnamon.

Colour

It can be uniform or not. When the drug is in powdered form, colour can give us an idea of the part of the plant we have before us. Green, for example, indicates that the drug comes from the leaves or aerated parts; brown from bark, stalks, or roots; white from rubbers or starches.

Taste

This test is applied only when specifically required for a given herbal material. It can be sweet, bitter, astringent, acid, salty, tart, nauseous, aromatic, *etc.*

Texture

This is tested on the untreated sample, by sight, lenses, water or reagents, in order to determine the cut surface, whether the material is soft or hard, fibrous, smooth, rough, granular, *etc.*

Macroscopic Analysis

Macroscopic characteristics are observed by sight or using a magnifying glass. Characteristics such as shape, size, hairiness, variegation, surface, fracture, section, thickness, and hardness of the whole plant or of parts of it are observed. Careful observation of the morphological characteristics of each organ allows us to identify the majority of drugs correctly.

- Stalks: type, section, leaf distribution.
- Leaves: form, variegation, hairs, texture.
- Inflorescences: Flower distribution, bracts.
- Flowers: calyx, corolla, stamens, carpels.
- Fruits: type, shape, and size.

- Seeds: size, colour, shape.
- Bark: colour, striations, wrinkles.
- Trunk: growth areas, vessels, medullar radios.
- Underground organs: shape, appearance, consistency.

Microscopic Analysis

Organoleptic and macroscopic characteristics are not enough for identification, so a microscopic examination is needed. Cellular content (grains of starch and alleurone, crystals, fats, and essences), cellular elements (vessels, tracheids, epidermic cells, stoma, parenchyma, collenchyma, suber, sclerenchyma, petrified cells, fiber, and hairs or trichomes) can be observed.

Microscopic characteristics and histological cross-sections are also important. We use a microscope to observe cellular elements such as hairs, vessels, sclereids, stoma, and acellular like crystals and grains of starch. Thus, the nature of the drugs can be established when macroscopic analysis is insufficient. This process is also used to eliminate the possibility of adulteration. In histological cross-sections, we can observe cellular content and structure.

THIN LAYER CHROMATOGRAPHY (TLC)

This is used in all types of control on natural products, having established itself as a very important analytical method in modern pharmacopoeias. It enables us to identify the components present in specific vegetable matter at a low cost and quickly. It can also be used as a semi-quantitative method. Its application is described in the European Pharmacopeia 8.0 [3]. The WHO Guidelines describe the classical method, the micromethod and the horizontal technique [2, 7 - 9]. Other related techniques are forced - flow planar chromatography (FFPC), rotation planar chromatography (RPC), over pressured - layer chromatography (OPLC), and electro - planar chromatography (EPC) [10].

QUALITATIVE AND QUANTITATIVE DETERMINATION OF ACTIVE COMPONENTS

Determination of Ash

Ash represents the mineral salt or inorganic matter content of the drug. In rigorous conditions, it is a constant and allows us to find falsifications due to other drugs, soils, or minerals. It gives an idea of the mineral matter contained in a plant, usually around 5%. There are three methods for determining it, which provide different information, total ash, acid insoluble ash and water soluble ash.

Total Ash

The total ash consists of carbonates, phosphates, silicates and silica. They represent the mineral salt or inorganic matter content of the drug. In rigorous conditions, it is a constant and allows us to find falsifications due to other drugs, soils, or minerals [3]. Measuring it is important, because mineral matter may be the cause of a pharmacological effect. Potassium salts, for example, have a diuretic effect on horsetail (*Equisetum arvense*), dandelion (*Taraxacum officinalis*), and Java Tea (*Ortosifon stamineus*). If the content is high, this can indicate pollution due to the addition of mineral matter or soil, especially in roots. The method determines the amount of material that remains after ignition. The residue corresponds to the ash derived from the plant itself (physiological ash) and to that of foreign matter (non-physiological matter), *e.g.* sand and soil [11].

Acid Insoluble Ash

Inorganic variables like calcium oxalate, silica, and carbonate content of the crude drug affects total Ash value. They are removed by treating with hydrochloric acid. Acid soluble ash is the residue left over.

Water-Soluble Ash

It is the difference in weight between the total ash and the residue after treatment of the total ash with water.

The complete methods are described in the [2, 7 - 9] WHO Guidelines and the International Pharmacopoeias [3 - 6].

Water and Volatile Matter

By water content we mean the water freely contained in vegetable matter. For good quality conservation, this should be less than 10% to prevent enzymatic processes, and to express the value of active ingredients compared with dry matter. There are three ways of determining this.

Gravimetric or Loss on Drying Method

An exact quantity of dry matter is weighed, pulverised or broken in pieces, and placed in an oven at about 110 °C, and then weighed every 30 minutes until a constant weight is achieved. The difference between the initial and final weight is the water content or apparent humidity. Volatile substances may be lost, so this method cannot always be used.

Volumetric or Azeotropic Method

Water content is determined by azeotropic drag. A specific quantity of vegetable matter is placed in a balloon flask with benzene, toluene or xylene, and is distilled in a closed circuit. The water steam condenses, and as it is immiscible, forms a separate and visible phase, with which volume is established.

Karl-Fisher Method

This is very useful in samples with low moisture content, and is based on the quantitative reaction of water and sulphur dioxide and iodine in an anhydrous atmosphere in presence of a base. Iodine and sulphurous anhydride react in the presence of water, forming sulphuric and iodhydric acid.

Extractable Matter

When no suitable methods for evaluating active ingredients in a drug by means of physical or chemical tests are available, the amount of extractable matter is measured using water, alcohol, ether, or organic solvents, and it shows the amount of active principles that can be obtained by hot extraction and/or cold maceration.

Among the processes of extraction for different phytochemicals, new extraction technologies which eliminate the need for organic solvents are microwave assisted extraction (MAE), supercritical fluid extraction (SFE) and accelerated solvent extraction (ASE) or pressurized liquid extraction (PLE) [12].

Pressurized Liquid Extraction (PLE) is commonly used for the extraction of persistent organic pollutants in environmental samples. The technique uses organic solvent at high temperature and pressure, which drastically improves the speed of the extraction process. The parameters that have a significant effect on the extraction efficiencies of marker compounds in botanical and herbal preparations by PLE are the time for extraction/volume of solvent, applied temperature and the nature of solvent used [12].

Pressurized Hot Water Extraction (PHWE) has been used for the extraction of essential oil components from plant materials to reduce the use of organic solvent [12].

Extraction by Centrifugation

Extracts and oils obtained by these process present higher aromatic qualities than those obtained using steam distillation. This is not a thermal process, thus natural anti-oxidant properties are more stable. Even so, internal friction of the prime matter causes an uncontrollable increase in temperature which may mean a thermal degradation and a darkening of the extract. This requires additional, very expensive, purification equipment, which raises the final cost of the product.

Supercritical Fluids Extraction (SFE)

Caffeine from coffee, cholesterol-free foods, pure natural anti-oxidants, can be achieved using super-critical fluids.

A super-critical fluid is a substance, mixture, or element, which under certain operative conditions of pressure and temperature, and thanks to mechanical operations, is above its critical point but below the pressure needed to condense it into a solid. Extraction *via* super-critical fluids is more environmentally friendly than conventional methods of extraction, because it uses gases such as CO_2 at high pressure, in a liquid or super-critical state, instead of chlorinated solvents producing toxic waste. CO_2 is what we call an ecological solvent (normally referred to as a green solvent). It is not toxic, it does not pollute, it is not inflammable, it is economic, easily recycled, and therefore, is not an environmental problem in terms of waste management. The following aspects of super-critical fluids stand out:

- They are powerful solvents and have a great capacity of penetration in solids, which allows a rapid and almost complete exhaustion of extractable solids.
- They can easily be completely separated from extracts, simply by modifying pressure or temperature, up to the point, if necessary, that the fluid becomes a gas.

The main drawback is the time of extraction, which is usually long. In fact, in some cases, it can take as much as 24 hours. With normal fluids extraction can be speeded up by mechanical shaking, but this presents problems when using super-critical fluids, which limits industrial use. Applying high frequency ultrasound has been tried to speed up the process of extraction with super-critical fluids [13]. The process is already in use for the extraction of oils from whole and ground natural almonds at different particle sizes.

Extraction in a Solid Phase

Columns or cartridges capable of retaining active principles are employed, which are later extracted using a small volume of solvent.

Volatile Oils. Steam Distillation

Essential oils, the volatile part of active ingredients contained in a plant, are therefore obtained through techniques of distillation, where the active ingredients are volatilised and condensed. They are analysed using gas chromatography GC and mass spectrometry if necessary, after extraction *via* distillation by steaming at atmospheric pressure.

Distillation consists in separating components of a sample depending on the different pressure of the steam and boiling point. The boiling point is the temperature at which, at a certain pressure, a liquid becomes a steam, or the temperature at which the pressure of a liquid is equivalent to that of the gas around it. When distillation is carried out at atmospheric pressure, it is the equivalent of a column of 760 mm Hg. A reduction in pressure also reduces the boiling point, while an increase in pressure increases it.

A mixture of components has not got a boiling point, but a range. Different essential oils have a wide variety of compositions and boiling points, and distillation is carried out depending on them. Different oils have different boiling points, so their distillation takes place at a range of temperatures, which varies, usually between 150-300 °C.

The advantage of this method of extraction is the low original investment needed to acquire the equipment and accessories. It is a simple, versatile, flexible process, which enables work with large quantities of prime matter in each cycle, even without treating these in advance. The time of extraction does not change, but the yield does. The drawbacks are that thermal degradation occurs in the essential oils obtained, such as oxidisation, hydrolysis, and olygomeryzation.

Physical Measurements in Essential Oils

Aroma, specific weight, refractory index, optical deviation (rotational capacity), solubility in alcohol-water mixtures (reduced alcohols).

Essential Oil Refractory Index

This is given by the relation between the speed of light in air, and the speed of the substance being tested. It is measured using Abbe's refractometer [14]. It is a useful measure for establishing the purity of essential oils, as their values appear in many pharmacopoeias.

Essential Oil Rotational Capacity

The rotational capacity of a liquid is the angle of rotation of its plane of light polarisation when a ray of polarised light is passed through a sample of said liquid. This rotation may be clockwise or anticlockwise.

Essential Oil Chemical Measurements

Free acidity index. Saponification index. Ester indexes. Aldehyde and ketone measures. Phenylhydrazone formation. Oxyme formation. Semi-carbazone formation. Bisulphite method. Acetyl index. Chromatographic Fingerprint. Spectroscopic methods.

Bitterness Value

Many plants contain bitter compounds that activate hunger and digestion, and excite the tongue taste papilla. As a reflex reaction, stomach motility increases helping secretion of gastric juices. Bitter value is determined by taste, comparing a solution of the drug extract with a solution of quinine hydrochloride (1g in 2000 ml of drinking water).

Haemolitic Acivity

Some plants produce active compounds called saponins, which produce foam in water solutions, and are naturally tensoactive. Many have haemolytic properties (disintegration of erythrocites), and are highly toxic when injected into the blood. This is called haemolysis. It is determined by comparing a solution of the drug extract with a solution of saponin (haemolytic activity of 1000 units per 1g).

Tannins

Tannins are polyphenolic hydro-soluble substances of vegetable origin. Apart from reacting in the classic phenol way, they precipitate gelatine, alkaloid salts, and heavy metals. They are determined by colouration and precipitation reactions, explained below.

Swelling Index

This is determined in drugs with mucilage. It is defined as the volume of water (ml) taken up by 1 g of the drug after swelling due to being in contact with an aqueous liquid. Agar, for example, is > 15, flax whole seeds > 4 agar powdered seeds > 4.5.

Phytosanitary Residues

These are measured by TLC, GC, or HPLC (Annex 3) [8].

Microbiological Contamination

In general, the plant is contaminated at the moment of collection, which diminishes yield at the time of extraction. Nevertheless, the process of manipulating the plant (grinding, packing, storage, *etc.*...) can cause additional contamination. With certain exceptions, a maximum of 106 bacteria and 105 yeast is allowed per gram (Annex 5) [8].

Aflatoxins

Aflatoxins are secondary metabolites of *Aspergillus flavus* and *A. parasiticus*. They are found in grain and in herb medicines when humid conditions are given. They are toxic, presenting carcinogenic, teratogenic and mutagenic properties. The analytical methods for them are HPLC - FLD with derivatization by trifluoroacetic acid (TFA) and Kobra cell (bromination) [15]. The method recommended by the WHO Guidelines for AF (B1, B2, G1 and G2) is HPLC using fluorescence detection, which does not require the use of toxic solvents such as chloroform (Annex 4) [8].

Metals and Nutrients

Metals, minerals and macro and micronutrients such as K, Na, Mg, Ca, Fe, Cu, Zn and Mn, are analysed from complete ashes, in 1g samples, calcined at 400 °C for 24 hours and dissolved in chlorohydric acid (1:1) and atomic absorption spectrophotometry (Annex 3) [8]. For toxic metals, such as Hg, Cr, Pb, pre-established criteria are accepted for foodstuffs of vegetable origin.

Radioactive Contamination

European regulations [3] allow a maximum level of 600 beq/kg, although this refers to food products and not specifically to medicinal plants.

Colouration and Precipitation Reactions

These are simple reactions specific to an active principle or characteristic component of a drug, which serve to identify it. They are used for alkaloids, glycosides, saponins, flavonoids, tannins, *etc.* These tests are not always applicable, due to the interference of other substances present in the drug, but they are very simple and quick. In general, they are done directly on a plant extract, usually of alcohol extraction.

For example, the appearance of a red colour in a plant drug when dissolved in an ammonia solution is due to its anthraquinone content. Similarly, a purple colouration when KOH is added to the residue of evaporating the alcoholic extract of a drug indicates the presence of furochromones. Cyanogenic glycosides are determined using pycroside paper, which turns reddish. This differentiates the sweet almond variety from the bitter one. Liebermann's reaction, specific to steroids, gives us a bluish-greenish shade when they are present in a plant drug.

Fluorescent Reactions

The rhubarb root (*Rheum palmatum, R. officinale*), for example, can be distinguished from a falsification using the root of the rapontic (*R. rhaponticum*) because the latter produces a blue-violet fluorescence at 366 nm, due to the fact that it contains raponticine, a substance absent in the official drug.

Chromatographic Fingerprint

There is a common practice among natural product analysts to select one or more compounds as either active or "markers" for purposes of identification and quality assessment. In Chromatographic fingerprint analysis multiple compounds in a single herbal can be identified. It utilizes chromatographic techniques to construct specific patterns of recognition for multiple compounds in herbal drugs. The entire pattern of compounds can then be evaluated to determine not only the absence or presence of desired markers or active ingredients but also the complete set of ratios of all detectable analytes [16].

Chromatographic techniques are based on separating the substances present in a complex mixture by putting it in contact with a mobile phase (liquid or gas) and another static one (solid or liquid) which is fixed. The substances migrate through the static phase dragged by the mobile one, at different speeds depending on affinity to or solubility in one phase or the other. Substances with a greater affinity to the static phase migrate more slowly, and those with greater affinity to the mobile phase, faster. If we adjust the chromatographic parameters, we manage to separate all the components. Three most common types are thin layer chromatography (TLC, described above), gas chromatography (GC), and high performance liquid chromatography (HPLC).

Gas Chromatography (GC)

Gas chromatography or gas-liquid chromatography GC is used with volatiles or volatile substances, such as essential oils. A gas is used as the mobile stage, and a liquid in a capillary column (of very small diameter and several metres in length) is used in the static stage. The column is rolled up inside the chromatograph. High temperatures are used so that the components of the mixture remain volatilised.

This is the most selective technique, and is recommended by the European Pharmacopoeia as the standard method for analysing them. It allows us to separate and identify essential oils, camphor, vegetable acids, certain alkaloids such as opium or tobacco, cannabis resin, and compound steroids such as sapogenins and cardiac glycosides. It is both quantitative and qualitative. This technique, coupled with Mass Spectrometry (GC-MS), allows us to obtain the mass spectrum of each component, whence we can find the molecular weight and structural information. There are databases with the mass spectrum of many components. More recently, chiral chromatographic columns have been developed for separating optically active components.

High Performance Liquid Chromatography (HPLC)

This technique enables to separate and identify molecules of very similar chemical structure and even isomers. It is used with non-volatile compounds, such as glycosides, alkaloids, lipids, steroids, carbohydrates, flavonoids, and thermo-labile substances like proteins and vitamins. Both the mobile and static phases are liquid, of different polarity, so the separation of substances occurs depending on this parameter. It is carried out at room temperature, but the mobile phases are pressure-pumped.

The challenge is to develop methods using columns with a smaller inner diameter and the right mobile phase that can be readily adapted to mass spectrometry. The most common mode of detection remains ultraviolet detection. Gradient elution HPLC with reversed phase columns is applied for the analysis of multiple constituents present in medicinal plants and herbal preparations [12].

Capillary Electrophoresis

Capillary electrophoresis (CE) is an alternative to HPLC in the analysis of polar and thermolabile compounds. Capillary Zone Electrophoresis (CZE) is based on the differences in the electrophoretic mobilities resulting in different velocities of migration of ionic species in the electrophoretic buffer in the capillary. Micellar Electrokinetic Capillary Chromatography (MEKC) is based on solute partitioning between the micellar phase and the solution phase. Capillary Isoelectric Focusing Electrophoresis (CIEF) has been used for the separation of natural products [12].

Spectrophotometric Methods

Many active ingredients can be chemically typified using the gas chromatography and mass spectrum data, as we explained above. However, when doubts as to character arise, we resort to spectral methods such as infra-red (IR), ultraviolet (UV), and nuclear magnetic resonance (NMR).

Near Infrared Spectroscopy (NIR)

The NIR spectrophotometry enables detecting the presence of hydroxile and carbonile, aromatic rings, double bonds C=C, *etc*... To establish the spectrum, we need only to place a drop of the component in a NaCl cell and put it in the spectrophometer. The NIR spectrum of a molecule is like its ID card, and it can be compared against a spectrum database.

Ultra Violet Spectrospcopy (UV)

The ultraviolet (UV) spectrophotometry allows recognising functional and chromophore groups (chemical groups which absorb UV). The lemon tree oil, for example, can absorb a maximum of 262 nm. On the whole, ultraviolet spectrophotometry has limited applications in the study of most terpenic essential oils, as few terpens contain chromophore groups.

Nevertheless, carotenoid components or components with oxygenated heterocyclic nuclei present in the non-volatile part of citric essential oils (coumarins, substituted furanocoumarins and polymethoxiflavones) give these substances a characteristic behaviour under UV.

This characteristic has been used to fine tone methods which enable the assessing of quality and authenticity, identifying the geographical origin of a sample, and the technology used for its extraction, or the season when the oil was produced.

Nuclear Magnetic Resonance Spectroscopy (NMR)

Thanks to developments in Nuclear Magnetic Resonance Spectroscopy (NMR) there are databases of spectra, especially C^{13} (^{13}C-NMR) for the most common monoterpens and sesquiterpens. Nuclear magnetic resonance spectroscopy (NMR) is only used in a few cases in pharmacopoeias for identification purposes, especially for large molecules. Examples are heparin, goserelin and buserelin in the Ph. Eur [3], and heparin, enoxaparin and oxytocin in the US Ph [4].

Mass Spectrometry (MS)

In connection with the study of essential oils, the combination of gas chromatography GC with mass spectrometry (GC-MS) is the one that has received most attention since its inception. The combined GC-MS technology allows us to obtain the mass spectrum of each component separately by GC. We can get structural information and obtain molecular weight. There are databases with mass spectra data for the chemical identification of many components of essential oils and other types of substance.

Hyphenated Techniques

The qualitative analysis or structure elucidation of the chemical components by simple techniques is not always possible. Chromatographic fingerprints of natural products utilize chromatographic hyphenated techniques that combine chromatography and spectrometry techniques to construct specific patterns for recognition of multiple compounds [12, 17 - 21]. These combined techniques are employed to perform on line composition and structural analyses of botanicals (Table **1**).

Table 1. Hyphenated Techniques.

Gas Chromatography	Mass Spectrometry	GC/MS
High Resolution Gas Chromatography	Mass Spectrometry	HRGC/MS
High Performance Liquid Chromatography	Mass Spectrometry	HPLC/MS
	Diode Array Detection	HPLC/DAD
	Tandem Mass Spectrometry	HPLC/MS/MS
	Nuclear Magnetic Resonance	HPLC/NMR
	Strong Anion Exchange	SAX-HPLC
	Reversed Phase Ion Pairing	RP-IPC-HPLC
Capillary Electrophoresis	Diode Array Detection	CE/DAD
	Mass Spectrometry	CE/MS
Mass Spectrometry	Mass Spectrometry	MS/MS
High Speed Counter Current Chromatography		HSCCC
Low Pressure Size Exclusion Chromatography		SEC

HPLC/MS is used with thermally labile molecules. The most common modes of ionization in HPLC/MS are Electrospray Ionization (ESI) and Atmospheric Pressure Chemical Ionization (APCI). Single quadruple, triple quadruple, ion-trap, time-of–flight, quadruple time-of-flight (Q-TOF) and others mass analyzers are used. However, these methods are still limited to conditions suitable for MS and standard availability [12].

CONFLICT OF INTEREST

The authors declare no conflict of interest, financial or otherwise.

ACKNOWLEDGEMENTS

Declared none.

REFERENCES

[1] Dušková E, Dušek K, Indrák P, Smékalová K. Postharvest changes in essential oil content and quality of lavender flowers. Ind Crops Prod 2016; 79(January): 225-31. [http://dx.doi.org/10.1016/j.indcrop.2015.11.007]

[2] World Health Organization. Quality control methods for herbal materials. 2011.

[3] European Directorate for the Quality of Medicines & Healthcare EDQM. Strasbourg: Council of Europe 2010.Ph Eur 8th Ed.

[4] The United States Pharmacopeia. The United States Pharmacopeial Convention. Rockville, MD 2014.

[5] The Japanese Pharmacopoeia. 16th ed., Tokyo: Ministry of Health, Labor and Welfare 2011.

[6] Indian Pharmacopoeia. Vol 3. New Delhi: Government of India, Ministry of Health and Family Welfare 2010; pp. 2467-72.

[7] World Health Organization. General Guidelines for Methodologies on Research on Traditional Medicine. Geneva: WHO Library Cataloguing-in-Publication Data 2000.

[8] World Health Organization. 2007.WHO guidelines for assessing quality of herbal medicines with reference to contaminants and residues.

[9] World Health Organization. WHO guidelines on good manufacturing practices (GMP) for herbal medicines. WHO Library Cataloguing-in-Publication Data 2007.

[10] Nyiredy S. Progress in forced-flow planar chromatography. J Chromatogr A 2003; 1000(1-2): 985-99. [http://dx.doi.org/10.1016/S0021-9673(03)00308-X] [PMID: 12877209]

[11] Pradhan N, Gavali J, Waghmare N. WHO (World Health Organization) guidelines for standardization of herbal drugs. IAMJ 2015; 3(8): 2238-43.

[12] Ong ES. Extraction methods and chemical standardization of botanicals and herbal preparations. J Chromatogr B Analyt Technol Biomed Life Sci 2004; 812(1-2): 23-33. [http://dx.doi.org/10.1016/S1570-0232(04)00647-6] [PMID: 15556486]

[13] Riera Franco de Sarabia E, Gallego-Juárez JA, Montoya Vitini F, *et al.* Separation or extraction method using supercritical fluid assisted by high-intensity ultasound. European Patent 2003.EP1547 679 A1

[14] Fellmuth B, Gaiser Ch, Fischer J. Determination of the Boltzmann constant — status and prospects Published 21 August 2006. IOP Publishing Ltd. Meas Sci Technol 2006; 17(10)

[15] Donghun L, Jihye L, Kwang-Geun L. Analysis of aflatoxins in herbal medicine and health functional foods. Food Control 2015; 48(February): 33-6.

[16] Xie P, Chen S, Liang YZ, Wang X, Tian R, Upton R. Chromatographic fingerprint analysis--a rational approach for quality assessment of traditional Chinese herbal medicine. J Chromatogr A 2006; 1112(1-2): 171-80. [http://dx.doi.org/10.1016/j.chroma.2005.12.091] [PMID: 16472540]

[17] Liu S, Liang Y-Z, Liu HT. Chemometrics applied to quality control and metabolomics for traditional Chinese medicines. J Chromatogr B Analyt Technol Biomed Life Sci 2016; 1015-1016(15): 82-91. [http://dx.doi.org/10.1016/j.jchromb.2016.02.011] [PMID: 26901849]

[18] Koh HL, Woo SO. Chinese proprietary medicine in Singapore: regulatory control of toxic heavy metals and undeclared drugs. Drug Saf 2000; 23(5): 351-62. [http://dx.doi.org/10.2165/00002018-200023050-00001] [PMID: 11085343]

[19] Cai Z, Lee FS, Wang XR, Yu WJ. J. Mass Spectrom. 37 (2002) 1013. [69] J. Sherma. J AOAC Int 2003; 86: 873. [PMID: 14632386]

[20] Sherma J. High-performance liquid chromatography/mass spectrometry analysis of botanical medicines and dietary supplements: a review. J AOAC Int 2003; 86(5): 873-81. [PMID: 14632386]

[21] Liang Y-Z, Xie P. Kelvin Chan. Quality control of herbal medicines. Biomedically Relevant Plant Components: Active Principles and Toxicants. Journal of Chromatography BVolume 2004; 812(1–2): 53-70. [http://dx.doi.org/10.1016/S1570-0232(04)00676-2]

CHAPTER 8

Medicinal and Aromatic Plants Sustainable Assessment

Silvia C. Guillén[1,*], M. Paz Arraiza[2] and Miguel A. Sarmiento[3]

[1] *Ecology Department, Agronomy Faculty, National University of Tucumán, Argentina*

[2] *Department of Forestry and Environmental Engineering and Management, Technical University of Madrid, Madrid, Spain*

[3] *Faculty of Forestry, National University of Santiago del Estero, Argentina.*

Abstract: Sustainable agriculture is based in the conservation of environmental and productive resources considering the socio-economical dimensions. The cultivation, production and collection of medicinal and aromatic plants (MAPs) are made up by a sector of the agriculture that should follow the concepts of sustainability. Our team researches in the area of sustainability, having developed agro-ecosystems and agro-industry valuation models. The objective was to develop evaluation tools with particular objectives: 1- an assessment model of lemon cultivation sustainability; 2- assessment tools in MAPs greenhouses and agro-industry; 3- introduction of principles for sustainable collection of wild species. Methodologies used by international organizations, like the UN or OECD, and specific models like FESLM (FAO) were taken as a reference. We worked with the construction of a Minimum Sustainable Set of Indicators (MSI) for each case. Sustainable assessment model for lemon cultivation were obtained. MSI for greenhouse's MAPs cultivation and lemon and blueberries packing houses were developed. International principles for sustainable wild collection of MAPs were introduced. Considering the selected indicators, the results obtained are of those analysed systems that should check over and adjust their process, objectives and policies in the search of an improvement tendency towards sustainability. The harvest of wild species should start its way towards sustainability through the approach in the uses of standards.

Keywords: Blueberries packing houses, Environment, Greenhouses, International standards, Lemon agro-industry, Lemon cultivation, MAPs plants, Sustainable agriculture, Sustainable assessment, Sustainable indicators, Sustainable wild collection of MAPs.

* **Corresponding author Silvia C. Guillén:** Ecology Department Agronomy Faculty, National University of Tucuman Argentina; Tel/Fax: +54 9 381 512 8980; E-mail: sguillen@faz.unt.edu.ar

M. Paz Arraiza, A. González-Coloma, J. Burillo and C. Calderón-Guerrero (Eds.)

INTRODUCTION

Global and local environmental and socioeconomic problems became important at an international level and were subject of academic studies that increased in the 1970s. The report of the World Commission on Environment and Development (1986) established the beginning of a world environmental vision and started a new stage where the concepts of sustainability and sustainable development were built and spread.

Sustainable, broadly speaking refers to the possibility of humanity not to compromise future life, strongly supporting the idea of the need to preserve our resources [1]. This implies having understood that the world is limited. Scientific studies and major international organizations actions have encouraged the development of sustainability concept; however it is not yet built-in to the population so it is considered to be a current subject for study, investigation and spreading. It is a concept and a way of working which has been transferred into different fields of the human life, such as agriculture.

Sustainable agriculture is characterized by the preservation of natural resources and the search for a higher environmental compatibility by means of the use of local renewable resources along with appropriate self-sufficiency technologies. It is an all-embracing approach towards food, forage, fibre production that balances the environmental well-being, social equity and economic feasibility amongst all social groups including international communities and future generations [1]. This agriculture pretends the management and maintenance of stock resources, goods and services production, to be done in at least a constant level of productivity. Numerous papers establish that the success of sustainability in agricultural activities is based in the way these systems get closer to natural processes [2].

Cultivation, production and harvesting of medicinal and aromatic plants (MAPs) constitute an agricultural sector that should follow the concepts of sustainability. It is important to remember that it merges social, environmental and economic objectives [3] which currently join institutional and political subjects.

The resources for collection and production of MAPs are part of complex systems with unknown relationships, where the use of one has an effect on other resources.

This, added to social dimension, generates usage and exploitation conflicts whose solutions require a holistic approach with the participation of different participants involved. Hence, in the analyses of the productive systems including the agricultural ecosystems, not only inner elements and variables are involved, but also outer factors. The principles and guidelines for the sustainable management of the agricultural ecosystems are the foundation for rural and agro-industrial sector administration, being many of them part of the current regulations that govern.

To find out if a production or collection system is sustainable, an assessment should be done. It has the questionable aspects such as the possibility to gauge all the factors involved. This complexity is observed when we have to give value to natural and cultural resources. However, this is not an obstacle to evaluate sustainable tendency by means of checking lists, indexes, and models based on the use of indicators within an interdisciplinary frame. The evaluation is always comparative and based on a simultaneous analysis of a system with an alternative one, or the analysis of one system through-out time.

Within the backgrounds in sustainable assessment using indicators, the Checking List developed by the Commission on Sustainable Development (CSD) of the United Nations stands out as a referent. It includes four categories of indicators (economical, environmental, social and institutional). Frames were developed for sustainable assessment that contemplates a theoretical analysis such as: Pressure-State-Response (PER) [4] and CSD Theme Indicator Framework [5]. In general terms the PER frame is based in the fact that human activities put the environment on pressure, causing changes in their state itself. The specific answers of society might reduce or mitigate these pressures. In the assessment of sustainability in agriculture, two frames stand out: FESLM (Framework for Evaluation Sustainable Land Management) [6] and MESMIS (Assessment Framework of Systems Management with Sustainable Indicators) [7].

The MESMIS frame is a project developed by a multidisciplinary team from Mexico that promotes a systematical, participatory, flexible, interdisciplinary and cyclical assessment process. The application of this frame has been oriented towards the design and assessment of agro-ecological innovations at a local scale. Apart from the systems discussed above, local particular models of application have been conducted, such as the development for the subtropical productive system assessment with a multi-criterion and integration analysis [8, 9].

All the sustainable assessment systems mentioned above are based on the use of indicators, which allow quantifying and simplifying complex phenomena. They transfer the concept of sustainability into numeric terms, descriptive measures, signs and tendencies. Not all the obtained system data should be considered indicators, as they must gather desirable characteristic such as: cost, reliability, importance, reproducibility and veracity in a way to be credible and understandable for the community. The most important functions of an indicator are: estimate conditions and tendencies of the systems compare through time-space and provide information in short notice.

The conservation and the sustainable and profitable use of biodiversity is an aim of sustainable agriculture and research. The particular case of MAPs represents a specific group of productive activities that men execute, where genetic diversity has a fundamental role. In addition to the reference to biodiversity, and the topics mentioned above, the need to search MAPs productive systems, that are environmental, social and economic sustainable, is highlighted.

When analysing MAPs in connection to sustainability, it is important to differentiate at least two supply systems: species cultivation in agricultural ecosystems and wild species collection. In the assessment of cultivated MAPs productive systems, different methods from those mentioned above can be applied as well as to develop specific tools in accordance to agro ecological regions, cultivated species, cultivation systems, environmental, social and economic objectives, amongst many other factors. In relation to wild MAPs collection, it is fundamental to consider that these are natural resources offered by native or naturalized flora of each eco region and its exploitation explains the challenge of natural patrimony appraisal and the biological variety conservation, generally associated with economic resources that are essential for harvesting communities. The MAPs exploitation in this scenario requires sustainable management so as not to make up another factor towards the destruction of the biodiversity and the social structures that carry out the exploitation.

According to the International Union for Conservation of Nature (UICN), there are between 50,000 and 80,000 MAPs used worldwide [10]. These species have their ecological and social role as they are in some places, the base of rural community economies [11, 12]. In spite of its importance, it is considered that approximately 15,000 species are threatened with extinction [10]. Among the main causes of this situation, the IUCN cites the over-exploitation and the unsustainable agricultural and forest practices, as well as deforestation, habitat destruction, pollution, introduction of exotic species and climate change. In the case of 44 species of medicinal plants from India that were added to the Red List in 2015 because of excessive collection and habitat destruction [13]. It is estimated that in Argentina about 1,500 species are exploited due to their medicinal properties [14].

In the search for solutions to the previous issues and considering the necessity to unify the criteria of sustainable collection of wild species, IUCN, WWF Plant Specialists Groups, German Federal Agency for the Conservation of Nature (BfN), are developing an International Standard for Wild Sustainable Collection of Medicinal and Aromatic Plants (ISSC-MAP). The work of these organizations began with the Guidelines′ revision about the Conservation of Medicinal Plants during 2003-2006 [15].

These International Regulations for Wild Sustainable Collection of MAPs aim to provide guidance towards sustainability of eco systems in a set of principles and criterions that can be applied to the management of aromatic and medicinal plants and the eco systems that contain them. It could also serve as a basic tool for certification and audits of wild collection in the organic sector. The Standard Mission proposed is to assure the survival of medicinal and aromatic plants population in their habitat, respecting traditions, cultures and well- being of the participants.

Within the considerations for the conservation of wild species it is summarized that: a. the majority of the species are now and will continue to be wildly collected, b. this happens more in countries with low income and represents an additional income or maybe the only one, c. supports the preservation of traditional knowledge and cultural identity. In the proposed Standard, the main objectives were: provide in a principle frame and criteria that could be applied to the management of aromatic and medicinal species and their eco systems; provide a guide for the wild sustainable collection MAPs and the bases for audits and certification [15].

The regulations used, must be agreed and be part of a management plan along with vigilance system that observes the changes in vegetable communities. So as not to produce negative changes such as the decrease of collected species due to excessive harvest, changes in the age, structure and dynamic of population. Positive changes can be promoted by the recovery of a threatened species, and the rejuvenation of the population.

The general scheme of ISSC-MAP considers Ecologic Principles such as maintain wild medicinal plants resources and minimize impacts; Social Principles like performance of law and regulations and respect to customs and collective rights; Socio Economic Principles apply responsible practice of management and business. The development of this Standard has different challenges due to the existence of several difficulties in the collection systems, multiple and easily differentiated cultural issues, and to the eco system diversity. Because of these, for each particular case as in the agricultural eco systems, the adjustment of objectives and methodologies to sustainable assessment are needed. These regulations are designed to help those who are involved in harvesting, trade, derivative manufacture, and the selling of wild MAPs collection [15, 16].

The Argentinian northwest is an eco-region where sub regions with different economic flair converge. It highlights agricultural production with their respective agribusinesses. The citrus industry is the second agro industrial activity of the province of Tucuman; it occupies a privileged place among world lemon exporters. This industry processes fresh fruit and essential oils products, peel and pectin. Another important production system is the fine fruit where bilberry production gained prominence in recent years from the opening of the US market and investments made by the companies involved. In contrast to these activities including field cultivation and industrialization, in the Argentinean northwest, MAPs are produced in greenhouses systems with significant changes in the environment [17].

Our team at the Ecology Department in the Faculty of Agronomy at the National University of Tucumán investigates in the field of sustainability applied to different areas of agriculture, agribusiness and the importance of sustainable management of wild species. The overall objective was to develop tools for sustainability assessment. The specific objectives in MAPs plants research were: (a). development of a model for assessing sustainability in crop species (lemon), (b). development of an assessment tool of sustainability in greenhouses cultivation system, (c). development of a tool to assess the sustainability of agribusiness in MAPs plants, and d. introduce methodologies for sustainable collection of wild species.

MATERIALS AND METHODS

The research was taken in: a. lemon cultivation, b. intensive productive system of aromatic, medicinal and condimental greenhouse, c. lemon and blueberry agro-industry and d. wild species management. In the research sustainable indicators were selected using worldwide referents such as Blue Book from the UN [5], OECD's PER Model [4], FAO's FESLM Model [6], MESMIS Model [7] among others, and at a national level Sarandón [18, 19], Viglizzo [20] and Blasco [21] methodologies were used.

The study was conducted in different areas in the province of Tucumán. The province covers an area of 22.524 km^2, in northwestern Argentina (Fig. **1**).

The study about sustainability of lemon cultivation was realized through soil studies, analyzing soil quality indicators with classical methodologies used by reference laboratories. The field work was done for five years in an agro-ecosystem located in the Yungas ecological region (26º55´30´´S, 65º22´30´´W), in the province of Tucuman in Argentina. Acknowledged chemical, physical and biological analytical methods were used: Chemical Fertility (labile carbon or active fraction, total organic carbon, total nitrogen, nitrates, *etc.*); Physical fertility (texture, bulk density, aggregate stability) and Biological Aspects (biological species richness, earthworm density, mulching or vegetal cover).

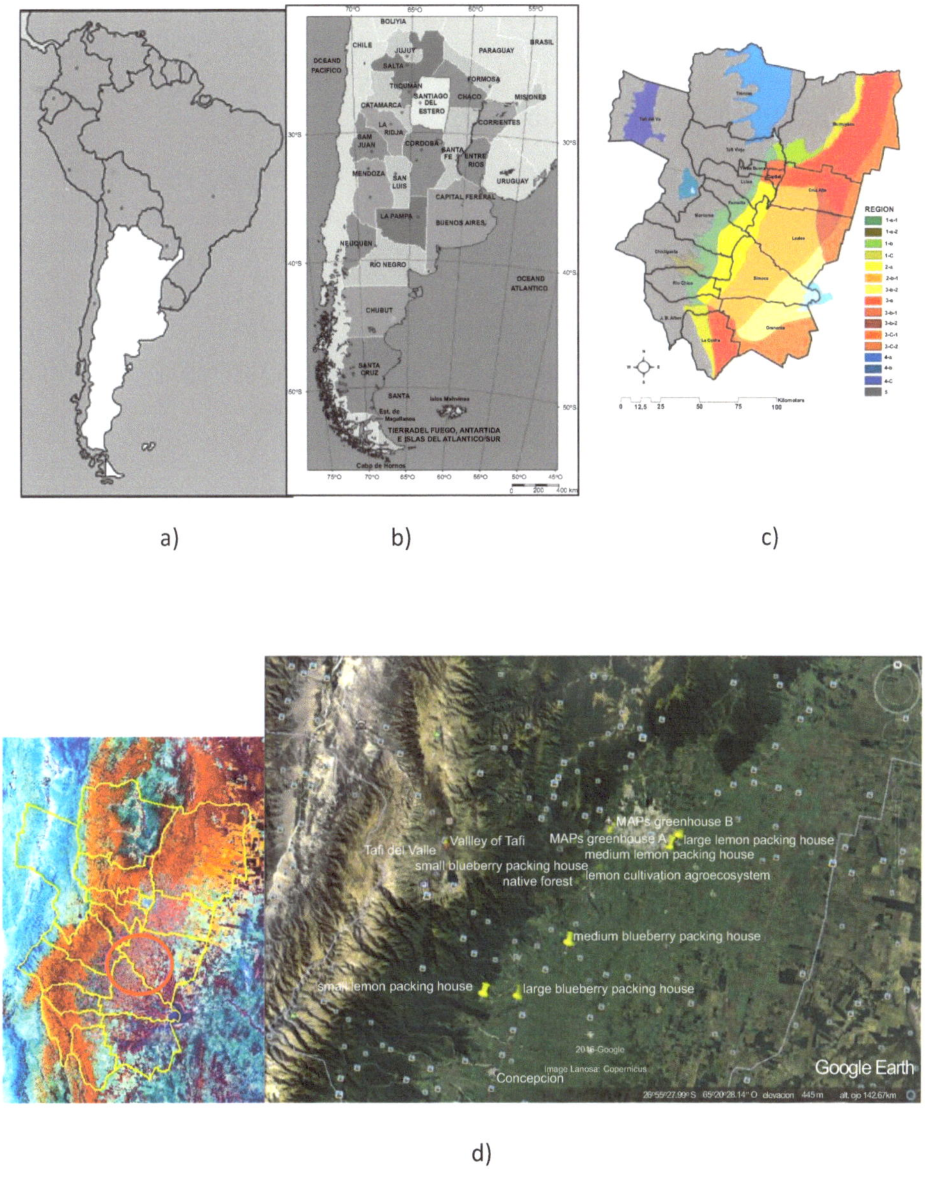

Fig. (1). (a) Location of Argentina in South America, (b) Location of Tucumán in Argentina, (c) Map of political division of the province of Tucumán with agro ecological regions (Yungas region in green color), (d) Location of the researched agro-ecosystems and agro-industries in the province of Tucumán, Argentina.

The data obtained were compared with the values of a secondary native Yungas´s forest soil located near the lemon agro-ecosystems analyzed (26º56´13´´S, 65º23´37´´W). Five hounded ninety data were collected and analyzed [8]. Partial Sustainability Indices (PSIs) were calculated using polar coordinates graphs where it was considered the relationship between the areas of the figure corresponding to the agricultural ecosystem and the area of the figure of the natural ecosystem of reference (Fig. **2a**). To carry out the calculation of the area and indices a formula [8, 22] was applied

$$IS = \frac{\left(\sum_{i=1}^{n} a_i\right)^2 - \sum_{i=1}^{n} a_i^2}{n(n-1)}, \quad \textbf{(1)}$$

Where *a* is the data of each parameter studied and *n* is the number of vectors or axes considered, according to the number of indicators. This formula was also applied by means of a common mathematical worksheet Microsoft® Excel® 2010 [8], to enable the calculation.

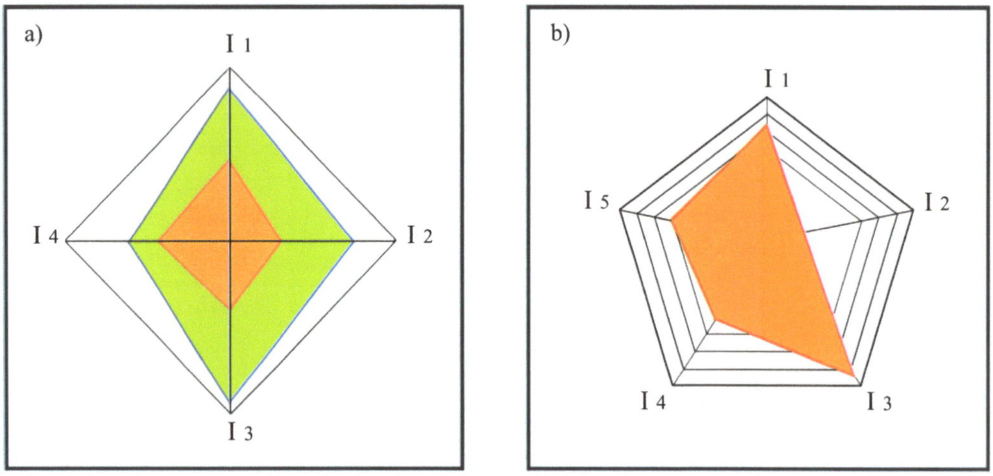

Fig. (2). (a) polar coordinates graph of agricultural ecosystem (orange color area) and the native forest as a reference system (green color area); (b) polar coordinates graph of a system under study with surveys (orange colored area) and the optimal values for each indicator (white color area).

Management made by the farmer affect the sustainability of the agro-ecosystem by which there was also a *Management Register Methods* of cultivation with the use of an *ad hoc* survey (Table **1**). This survey was generated in collaboration with different participants involved in the agricultural production studied. It was assessed with scales of 1 to 5, with 1 being the lowest and 5 the highest sustainability. Indicators were selected according to their flairs to build a Minimal Set of Indicators (MSI). PSIs were generated obtained by applying a formula (**1**) generated from polar coordinates graph (Fig. **2b**) [8].

Table 1. Scheme survey for the assessment of sustainability related to the management of soil vector in lemon cultivation.

Questionnaire for the assessment of sustainability related to soil management - Lemon Cultivation

A- Chemical management of the soil

1. Type of fertilization used

Options	
1. Chemical fertilizers applied in once	
2. Chemical fertilizers applied in divided doses	
3. Biological fertilizers applied in once	
4. Biological fertilizers applied in divided doses	
5. Green manures	

2. Amounts to be applied per hectare (Nitrogen)

Options	
1. $\geq$ 400 kg/hectare	
2. 300 $\leq$ 399 kg/hectare	
3. 200 $\leq$ 299 kg/hectare	
4. 100 $\leq$ 199 kg/hectare	
5. 0 $\leq$ 99 kg/hectare	

3. Adjustment of fertilization

Options	
1. Fertilization by unqualified references	
2. Fertilization by qualified general references	
3. Reference fertilization performance	
4. Fertilization according to a unique laboratory assessment	
5. Fertilization according to periodic laboratory assessment	

The researches in lemon and blueberry agro-industry were based, as well as work in agricultural ecosystems green housing, on the four sustainability pillars and the same methodology was used for sustainability assessment [21]. Three lemon packing houses were selected from the 37 enabled for export, with one of each category sorted by volume of daily production in large, medium and small agro-industry (I 26º50′47′′S, 65º08′24′′W; II 26º52′14′′S, 65º09′59′′W; III 27º11′20′′S, 65º37′56′′W respectively) (Fig. **1**). For the assessment of blueberry agro-industry sustainability, three packing houses were selected according to their productive capacity (I large 27º11′41′′S, 65º32′41′′W; II medium 27º04′44′′S, 65º25′04W and III small 26º55′23′′S, 65º20′30′′W) (Fig. **1**). Since blueberry fruit process is done without water, this resource analysis was not included in environmental indicators. Sustainability studies of housing MAPs were made in two representative agricultural ecosystems (A: 26º49′45′′S, 65º18′40′′W and B: 26º48′47′′, 65º17′51′′) located in the ecological region of Yungas in the province of Tucumán, Argentina (Fig. **1**). The principal MAPs and dye plants cultivated in both greenhouses were: mint, balm, rosemary, aloe vera, thyme and oregano.

In relation to the management of wild MAPs collection, management sustainability systems at a world level were investigated. Professionals were trained to disseminate methodologies and a primary analysis was done for the sustainable management of wild species in the agro ecological region Valley of Tafí (Argentina).

RESULTS AND DISCUSSION

The results in different cases discussed were:

Lemon Cultivation Sustainability Assessment

In lemon cultivation a Model of Sustainability Assessment in Subtropical Ecosystems was generated [8]. This included a number of results: 1. In relation to the soil study, the cultivation of lemon shows a significant deterioration in comparison to the system of reference, the native forest (Fig. **3**). 2. In relation to the selection of Indicators for the Assessment of Sustainability in subtropical cultivation, the Minimum Set of Indicators (MSI) resultant contemplates the Assessment of the physical, chemical and biological degradation with a number of 19 indicators. 3. The Model proposed through the soil management survey approaches the ecological principles to the understanding and development of Sustainable Agricultural Ecosystems (Fig. **3** and **4**). 4. The Model allows farmers to introduce the concept of sustainability, and to find means to achieve it, in their own agricultural ecosystems.

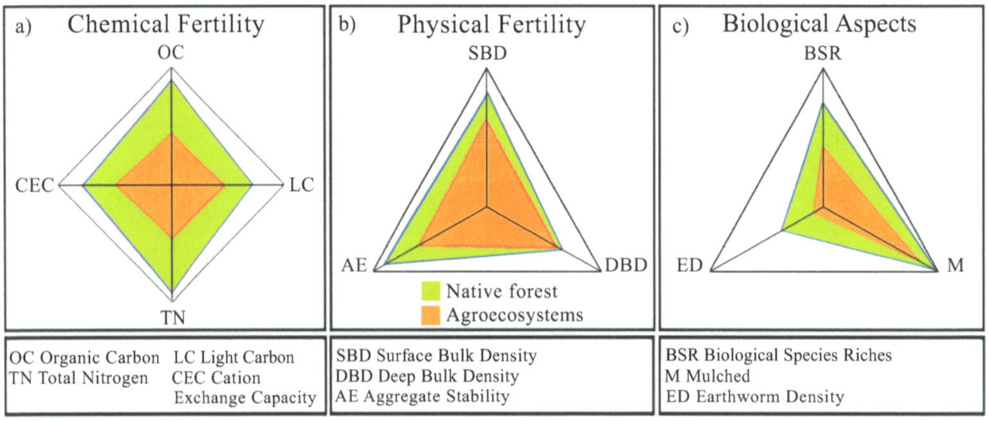

Fig. (3). Representation of the Sustainable Indicators in relation to chemical and physical fertility, and the biological aspects of the soil in lemon cultivation.

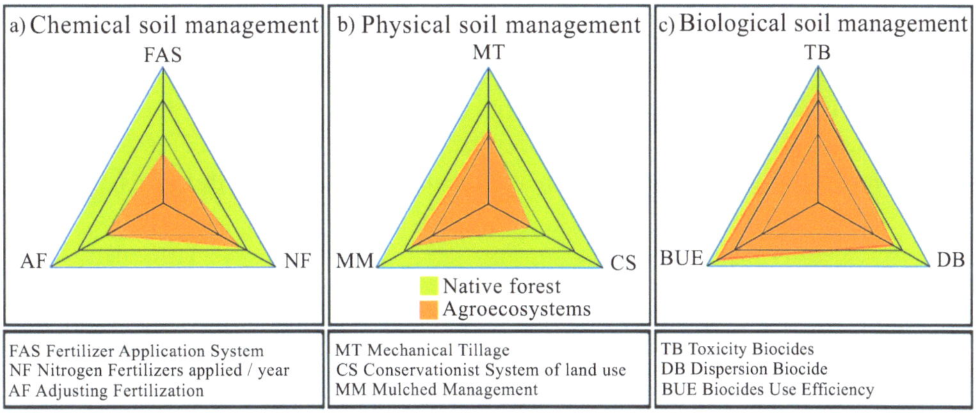

Fig. (4). Representation of the Sustainable Indicators in relation to chemical, physical and biocide soil management in lemon cultivation.

MAPs Sustainability Assessment in Greenhouses

A specific MSI was generated for these particular productive systems that could do a comparative analysis of sustainability. The values of Partial Environmental Sustainability Indices PESI obtained in both systems showed a significant difference, being the system with professional management (B) the one with the highest value marked by friendly environmentally application techniques (Fig. **5**) [24 - 26].

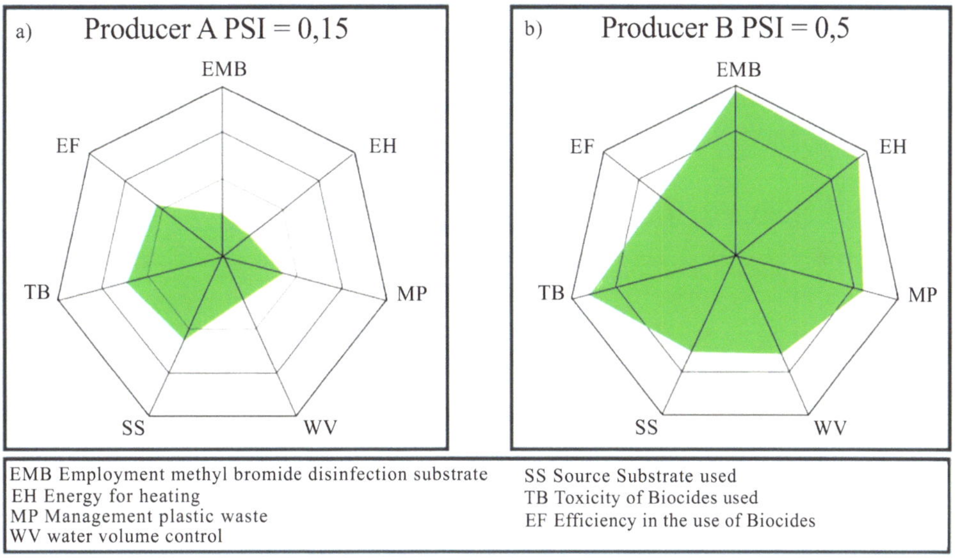

Fig. (5). Representation of the Environmental Sustainable Indicators in two MAPs greenhouses productive systems.

Lemon and Blueberries Sustainability Assessment in Packing Houses

With the data obtained from the surveys, graphs were made corresponding to the studied agro-industries differentiated by the four sustainable themes. As an example, the result of the graphic representation of indicators of environmental sustainability in lemon packing houses companies is expressed in Fig. (**6**) [23].

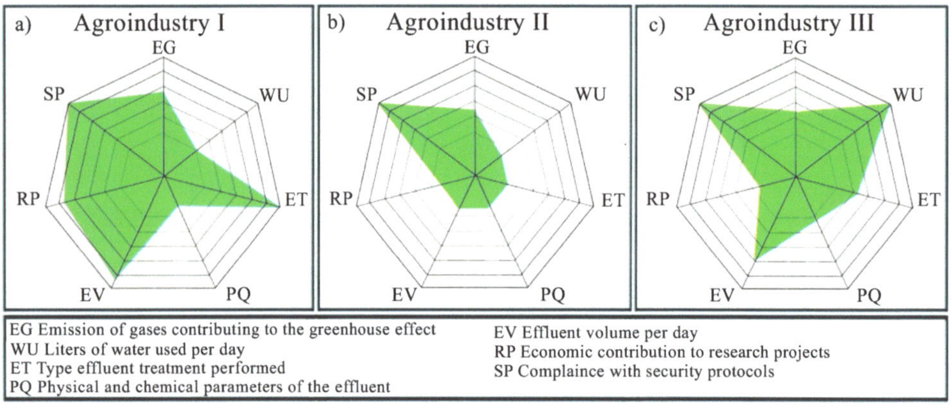

Fig. (6). Representation of the Environmental Sustainable Indicators in lemon fresh fruit packing houses.

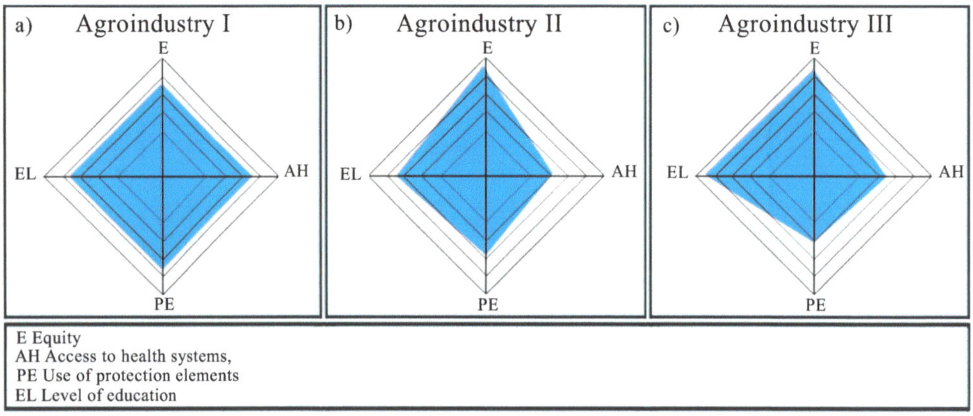

Fig. (7). Polar coordinates graphs generated from the surveys for the three blueberry packing houses, considering Social subjects.

In an environmental aspect the PSI were markedly different in the three packing houses. In relation to environmental sustainability, the agro-industry that processes most fruit presented a higher value, mainly due to effluent treatment and the quantity control of the water resource used in the industrial process.

In blueberry packing houses, the results indicated greater sustainability in the social and economic- institutional aspect (Figs. **7** & **8**) and ultimately the environmental aspect [27].

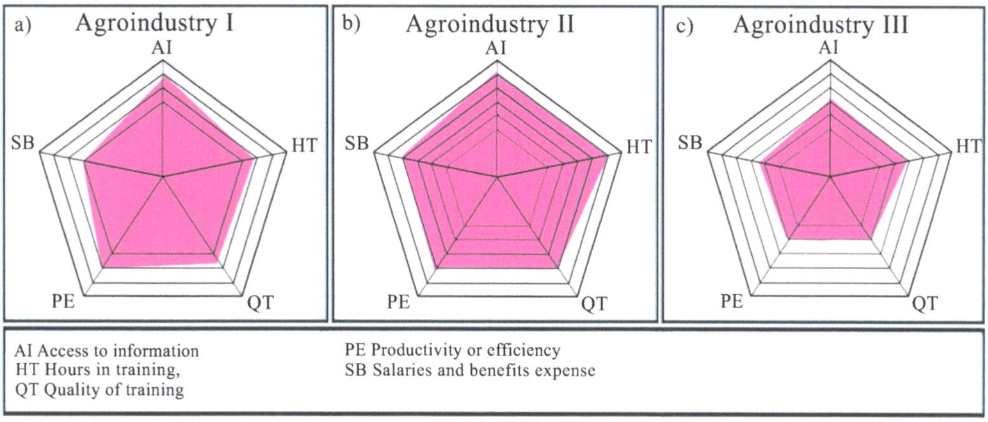

Fig. (8). Polar coordinates graphs generated from the surveys for the three blueberry packing houses, considering Institutional-Economic subjects.

Sustainable Management of MAPS Collection

In relation to the management of wild species the results were qualified professional trained in standards, sustainable management systems that are performed worldwide were disseminated and a first approach for sustainable management of wild species in the Valley of Tafí was carried out following Latin American researchers [7, 14, 16].

The methodology proposed in the Blue Book of the UN Sustainable Development Commission could be adapted for the assessment of sustainability in the several experiences made with MAPs. The methodology of generating specific surveys and adjusts them with the productive systems managers approach towards the PSR (pressure-state-response) model of the OECD and MESMIS Model.

CONCLUDING REMARKS

Of the results obtained in the different experiences, it can be concluded that the MSI of the indicators selected for the different studied cases, established a valid and simple application tool to assess sustainability. The applied methodology which includes indicators by topics and subtopics, obtained data from *ad hoc* surveys, and the incorporation of the four sustainable pillars, allowed a better comparative analysis of the condition of the different areas referred. Considering the selected indicators and given the results obtained it can be concluded that cultivation production systems and the analyzed agro-industries should review and adjust their processes, objectives and policies in search of an improvement in the trend towards sustainability. The wild MAPs collection should start its way towards sustainability by bringing the use of standards, which requires prior awareness on the part of the participants involved in the extraction and marketing chain.

CONFLICT OF INTEREST

The authors declare no conflict of interest, financial or otherwise.

ACKNOWLEDGEMENTS

Secretary of Science, Art and Technology Innovation of National University of Tucumán, Argentina.

REFERENCES

[1] Gliessman SR. Agroecology: The ecology of sustainable food systems. 2a ed., Boca Raton: CRC Press, Taylor & Francis Group 2007.

[2] Altieri MA. Bases y estrategias agroecológicas para una Agricultura Sustentable. Agroecología y Desarrollo. Revista de CLADES. Centro Latino Americano de Desarrollo Sustentable. Nº Especial 1995; 8/9: 21-30.

[3] Gallopín GC. La sostenibilidad ambiental del desarrollo en Argentina: tres futuros. Buenos Aires: CEPAL Comisión Naciones Unidas Serie Medio Ambiente y Desarrollo Nº 91 2004.

[4] Organization for Economic Cooperation and Development OECD. Core Set of Indicators for Environmental Performance Reviews: A synthesis report by the Group on the State of the Environment 1993. Available at: http://enrin.grida.no/htmls/armenia/soe2000/eng/oecdind.pdf

[5] United Nations Organization. Indicators of Sustainable Development: Guidelines and Methodologies 3rd ed.., 2007. Available at: http://www.un.org/esa/sustdev/natlinfo/indicators/guidelines.pdf

[6] Food and Agriculture Organization of the United Nations FAO FESLM. An international framework for evaluating sustainable land management 1993. Available at: https://www.mpl.ird.fr/crea/taller-colombia/FAO/AGLL/pdfdocs/feslm.pdf

[7] Masera O, Astier M, López-Ridaura Y. Sustentabilidad y manejo de Recursos Naturales. El marco de evaluación MESMIS. Distrito Federal: Mundi-Prensa, UNAM 1999.

[8] Guillén SC. Modelo de Valoración de la sostenibilidad en agroecosistemas subtropicales. Madrid: Universidad Politécnica de Madrid, Escuela Técnica Superior de Ingenieros de Montes 2008.

[9] Guillén SC. Quiénes y cómo evalúan la sostenibilidad. Una mirada desde los recursos naturales y el sector agrario 1a ed San Miguel de Tucumán: Facultad de Agronomía y Zootecnia UNT. 2009.

[10] Bentley R, Ed. Medicinal plants. London: Domvill-Fife Press 2010; pp. 23-46.

[11] Holt-Giménez E, Altieri MA. Agroecología, soberanía alimentaria y la nueva revolución verde. Univ Murcia, España Agroecología 2013; 8: 65-72.

[12] Fretes F. Plantas medicinales y aromáticas. Una alternativa de producción comercial UASID Available at: https://www.usaid.gov/sites/default/files/documents/1862/plantas_medicinales.pdf

[13] Unión Internacional para la Conservación de la Naturaleza UICN. Actualización de la Lista Roja de la UICN: éxitos de conservación ensombrecidos por màs declive de las especies 2015. Available at: https://www.iucn.org/es/content/actualizaci%C3%B3n-de-la-lista-roja-de-la-uicn-%C3%A-

xitos-de-conservaci%C3%B3n-ensombrecidos-por-m%C3%A1s

[14] Barboza GE, Cantero JJ, Nuñez C, Ariza Espinar L. Flora Medicinal de la Provincia de Córdoba (Argentina). Pteridófitas y Antófitas silvestres o naturalizadas. Córdoba: Gráficamente Editores 2006.

[15] Medicinal Plant Specialist Group Species Survival Commission IUCN – The World Conservation Union. International Standard for Sustainable Wild Collection of Medicinal and Aromatic Plants (ISSC-MAP) Version 10 Available at: http://www.floraweb.de/map-pro/Standard_Version1_0.pdf

[16] Monroy Ceseña MA. Diseño de criterios e indicadores de sustentabilidad para el aprovechamiento sustentable de la damiana silvestre (Turnera diffusa, Willd.) en la delegación El Carrizal, B. C. S. Universidad Autónoma De Baja California Sur 2012. Available at: http://biblio.uabcs.mx/tesis/TE2830.pdf

[17] IDEP. Tucumàn. Instituto de Desarrollo Productivo de Tucumàn. El limón en Tucumàn, Argentina y el mundo Informe Especial 2015. Available at: www.idep.gov.ar/admint/uploads/archivos/1387374562.pdf

[18] Sarandón SJ. El desarrollo y uso de indicadores para evaluar la sustentabilidad de los agroecosistemas. In: Sarandón SJ, Ed. Agroecología El camino hacia una agricultura sustentable. Buenos Aires: Ediciones Científicas Americanas 2002; pp. 393-414.

[19] Sarandon JS, Flores CC. Evaluación de la sustentabilidad en agroecosistemas: una propuesta metodológica 2010. Available at: http://campus.fca.uncu.edu.ar/pluginfile.php/22251/mod_resource/content/2/Bibl%202010%20Sarandon%20Flores,%20Indicadores,%20Rev%20Agroecologia.pdf

[20] Viglizzo EF, Frank F, Bernardos J, Buschiazzo DE, Cabo S. A rapid method for assessing the environmental performance of commercial farms in the Pampas of Argentina. Environ Monit Assess 2006; 117(1-3): 109-34.
[http://dx.doi.org/10.1007/s10661-006-7981-y] [PMID: 16917702]

[21] Blasco J. Indicadores para la empresa Madrid: Cuadernos de Sostenibilidad y Patrimonio Natural. España: Fundación Santander Central Hispano 2006.

[22] Martínez Santa-María C, Fernández Yuste J. Índices de alteración hidrológica en ecosistemas fluviales. Madrid: Centro de Estudios de Técnicas aplicadas Centro de publicaciones, Secretaria General Técnica, Ministerio de Fomento, CEDEX 2006.

[23] Canelada Lozzia MI. Desarrollo de Indicadores para evaluar la sostenibilidad en la agroindustria citrícola. San Miguel de Tucumán: Universidad Nacional de Tucumán, Facultad de Bioquímica, Química y Farmacia 2010.

[24] Canelada Lozzia MI, Guillén SC, Portas AM, Grancelli SM, Vidal PJ, Arraiza MP, *et al.* Desarrollo de indicadores de sostenibilidad para la producción de plantas ornamentales, aromáticas, medicinales y condimenticias bajo cubierta. Tucumàn, Argentina En: Libro de Ponencias: IX Encuentro E-ICES 9 Malatgüe; International Center for Earth Sciences, Comisión Nacional de Energía Atómica. 2013; p. 35.

[25] Canelada Lozzia MI, Guillén SC, Grancelli SM, Portas A, Vidal PJ, Reguera MC, *et al.* Cálculo de sostenibilidad ambiental en un sistema productivo de plantas ornamentales, florales, aromáticas, medicinales y condimenticias bajo cubierta. Tucumàn, Argentina Rev agron noroeste argent. 2014.

[26] Grancelli SM, Canelada Lozzia MI, Guillén SC, Juri S, Vidal PJ, Arraiza MP, *et al.* Análisis comparativo de la sostenibilidad ambiental en dos sistemas productivos de plantas ornamentales, florales, aromáticas, medicinales y condimenticias, bajo cubierta. Tucumàn Argentina En: Libro de Ponencias: X Encuentro E-ICES 10 Buenos Aires; International Center for Earth Sciences, Comisión Nacional de Energía Atómica. 2014; p. 50.

[27] Canelada Lozzia MI, Guillén SC, Grancelli SM, Juri S, Vidal PJ, Arraiza MP. Evaluación de la sostenibilidad en empaques de arándanos. Tucumán-Argentina. Rev Agron Noroeste Argent 2015; 35(2): 39-45.

CHAPTER 9

Economic Valuation of Environmental Services Provided by Medicinal, Aromatic and Dye Plants

Miguel A. Sarmiento[1,*] and **M. Paz Arraiza**[2]

[1] *Faculty of Forestry, National University of Santiago del Estero, Argentina*

[2] *Department of Forestry and Environmental Engineering and Management, Technical University of Madrid, Madrid, Spain*

Abstract: This chapter presents the results of studies carried out in Argentina that assessed people's willingness to pay (WTP) and willingness to accept (WTA) for useful plants conservation using the Contingent Valuation Methodology (CVM) and Avoided Costs Method (ACM). The WTP of residents in rural areas in Santiago del Estero, Argentina was researched. In this study residents expressed that medicinal, aromatic and dye plants are very important to them. In different parts of the province of Santiago del Estero, 125 families in six small towns were surveyed obtaining values of Consumer Surplus (CS) of $19.67 and Variation Compensatory (VC) of $22.156 (at an exchange rate US $1 equivalent to $3.20). Regarding dye plants, 35 artisans in these communities were also surveyed using the same method, and they expressed a willingness to pay (WTP) equivalent to $555.29 per month to preserve such plants. This tradition has been passed down over many generations, which ensures the availability of a continuous flow of environmental services. The results expressed in monetary terms that show the importance that environmental services provided by native forests for rural communities.

Keywords: Avoided Costs Method, Contingent Valuation Method, Dye and aromatic plants, Economic valuation, Medicinal, Native forests, NTFP, Rural area, Santiago del Estero, WTA, WTP.

INTRODUCTION

Non-Timber Forest Products (NTFPs) are all biological materials, other than timber, which are extracted from forests, other wooded land and trees outside forests that include products used as food and food additives (edible nuts, mushrooms, fruits, herbs, spices and condiments, aromatic plants), fibers (used in construction, furniture, clothing or utensils), resins, gums, and plant and animal products used for medicinal, cosmetic or cultural purpose for human use [1].

* **Corresponding author Miguel A. Sarmiento:** Forest Economics and Management, Faculty of Forestry, National University of Santiago del Estero, Argentina; Tel/Fax: +54 9 385 483 9264; E-mail: migui@unse.edu.ar

M. Paz Arraiza, A. González-Coloma, J. Burillo and C. Calderón-Guerrero (Eds.)

Forest is one of the main resources that provides people with goods and services. The potential offered by the vegetation of the Argentinean native forests in the use of plants for environmental, economic, aromatic, condiment, medicinal interests and herbal remedies is becoming more remarkable. Other authors [2] maintain the same approach as Tunisia.

The ecosystem services are components of nature, directly enjoyed, consumed, or used to yield human well-being [3]. Medicinal plant species constitute an important alternative to conventional medicine in a large number of developing countries, especially within poor communities that inhabit rural areas and lack access to health services [4]. Similarly, plants provide extra food supply for people living in poor regions.

However, knowledge of cultivation and maintenance of these useful plants in rural communities has decreased in recent decades due to the modernization of manufacturing processes of food, medicines, and other uses [4].

Price-based approaches to policy analysis require non-market valuation methods that are founded on theory primarily developed in the context of developed market societies, where people are accustomed to trading a wide range of goods and services for money [5].

Nonmarket valuation, and other economic techniques that emphasize exchange values over cultural and ecological values, have been subject to criticism regarding the inability of exchange values to represent the total value of an ecosystem [6].

Many of these forest goods and services do not have a market price to reflect the benefits they provide to society. While several goods such as timber have market prices, or are at least partially traded in the market (such as game for hunting), most forest services are generally non-priced public goods (leisure and recreation, watershed protection, biodiversity, *etc.*) [7]

The values depend on the complex interrelationship of these three subsystems: nature, economy and society. Moreover, according to Winkler [8], most valuation approaches are static or quasi-static. Especially in the context of the valuation of ecosystem goods and services, this is an unjustified oversimplification, as there are often trade-offs between the short-term and long-term use of nature. Thus, a suitable valuation method has to take into account irreversibility and long-term consequences. As a consequence, values are not static but dynamic and depend on the co-evolution of the three subsystems nature, economy and society [8].

Knowledge of the economic value is important for the optimal management of forests. Sarmiento [9] explained that knowledge of its true value can locate the different resources in an appropriate scale of importance that determines priorities for use and current and future consumption for both individuals and governments. Also, non-timber forest products, closely linked to environmental services, support local production of handicrafts, thus generating additional income to the household economy of the area. For example in Ecuador, Tinitana *et al.* [10] evaluated the following aspect on their study: 1) to create a novel list of medicinal plants sold at 33 traditional markets; 2) to establish medicinal plant use agreement among vendors with the Factor of Informant Consensus (FIC); and 3) to determine the most sold medicinal plant species using the Fidelity Level (FL).

Studies of indigenous and popular knowledge about natural resources are becoming increasingly important in defining strategies and actions for the conservation or recovery of forests. The key role of indigenous and local knowledge in biodiversity conservation and management has been consistently highlighted within many international directives [11]. The use of plants and plant products for different purposes such as food, wood, medicine, fiber, oil, fodder, aroma and ornament could be traced as far back as the beginning of human civilization [12].

In Argentina Morales and Ladio [13, 14] published the results of a study on the presence of plant complexes in the Mapuche medicinal and edible flora of Patagonia, Argentina.

Others studies on medicinal plant use have been conducted in several localities in Santiago del Estero, northern Argentina, by Díaz [15], Sarmiento *et al.* [16]. Regarding useful plants for food, Sarmiento *et al.* [17, 18] published studies on consumers' tastes and preferences. He also did research on the knowledge of native forest food products in urban populations. The properties of dye plants were valued by Sarmiento and Guerrero Maldonado [19]. These studies and research were carried out within a thematic network of medicinal, aromatic and dye plants in northern Argentina, coordinated by Arraiza *et al.* [20].

OBJECTIVES

The general objective of this study is to estimate the economic value of useful plants in different areas in Santiago del Estero, Argentina, as well as to evaluate rural people`s willingness to pay (WTP) and willingness to accept (WTA) for the sustainable management and maintenance of native forests using the Contingent Valuation Method (CVM) and Avoided Costs Method (ACM).

The specific objectives are: (i) to estimate the value of ecosystem services of medicinal, aromatic and dye plant uses among residents of rural areas; (ii) to evaluate their uses and cultural importance, and (iii) to identify rural people's knowledge level of such plants

MATERIALS AND METHODS

Materials

Study Area

The study was conducted in different rural areas in the province of Santiago del Estero. The province has a population of 874,006 inhabitants [21] and covers an area of 136.351 km^2, in northwestern Argentina (Fig. **1**). The study area lies between latitudes 20° to 30° S and longitudes 62° to 66° W. The inhabitants are mostly rural people with low incomes and far from major urban centers. Productive economic activities aimed at low-scale subsistence are predominant in most of these areas.

Social Aspects

One of the first studies of economic and environmental valuation of useful plants was carried out in Santos Lugares, Departmento Alberdi, Santiago del Estero. 17 out of 200 families that make up the Union of Small Farmers of North Salado located near Río Salado were surveyed [16].

Another study was developed in Guampacha, a town located in Sierra de Guasayán which occupies part of Departamento Choya and Guasayán. Guampacha covers an area of 60,000 ha and is located 63 km from the capital city. It has a population of 200 families, of which 30 were surveyed using 18 study variables [15].

South of Departmento Figueroa (in the central part of the province) the Municipal Commission of Villa Figueroa with 2,000 inhabitants (including nearby villages) is located. For the study of medicinal plants, surveys to 39 families were conducted in three nearby towns in southeast Villa Figueroa: San Felix, San Ignacio and Santa Maria. Their people are devoted to raising livestock (goats and cattle), rainfed agriculture (vegetables and maize) and crafts (leather and textile). Another major activity in the area is the production of charcoal [19].

Of the 15 craftswomen using dye plants registered in south Departmento Figueroa, three live in these three towns, with a total population of 39 families, 195 inhabitants. San Felix is the place with more families (18), followed by Santa

Maria (14) and San Ignacio (7) [22, 19]. These three towns are located about 8 kilometers from the east bank of Rio Salado and about 100 kilometers from Santiago del Estero capital city.

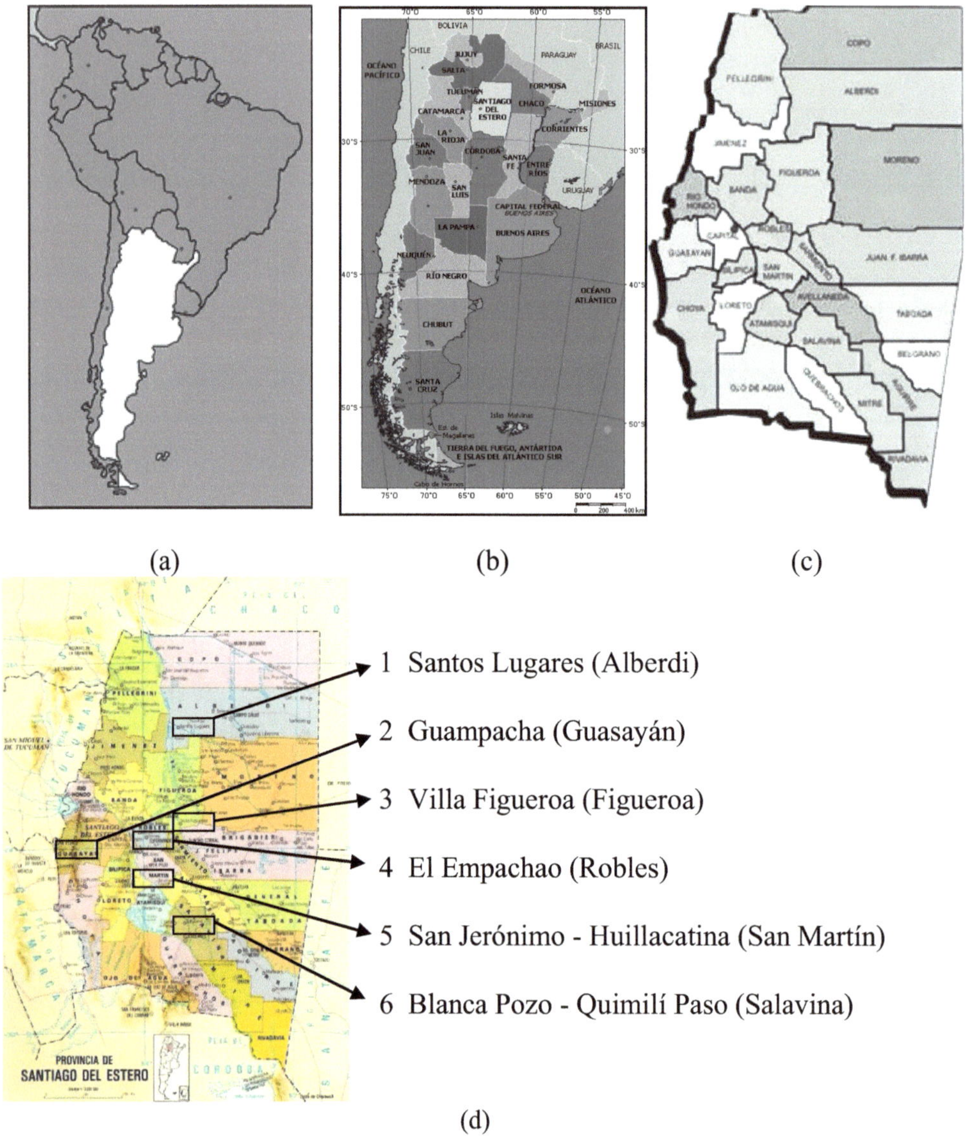

Fig. (1). (**a**) Location of Argentina in South America, (**b**) Location of Santiago del Estero in Argentina, (**c**) Map of political division of the province of Santiago del Estero, (**d**) Location of the researched towns in the province of Santiago del Estero, Argentina.

There is a project to build a water treatment plant, but currently houses (generally mudhuts) lack domestic water service, and drinking water is trucked in and stored in 10,000-litre deposits. Water for other uses, such as livestock and crops, are obtained through rainfall and by drilling water wells [22, 19].

The Municipal Commission of El Empachao is situated 72 km east of the capital city of Santiago del Estero, Fig. (**1**) in the western part of Departmento Sarmiento (very near the border with Departmento Robles). Perhaps this is one of the best connected of the towns surveyed, as most of its area is distributed along Provincial Route 89. 48 families live in this place, with an average of 5.5 members per family, totaling a population of 279.

The community Quimilí Paso is located in the eastern part of Departmento Salavina (close to the border with Departmento Avellaneda), about 250 km southeast of Santiago del Estero capital city and about 10 km northeast of Rio Dulce (it belongs to the alluvial plain Rio Dulce). It depends on the Municipal Commission of "La Paliza". The community consists of 24 families totaling 100 people (48 women and 52 men) [22]. The average number of people per family is 4.5, and the families interviewed ranged between 2 and 12 people. Of the total population 47% is in the age range of 0-19. From there the percentage decreases mainly because they (mostly women) migrate to other places in search for work.

Surveys were conducted personally and individually on a population accessed at random in the locations described, and trying to cover as many families as possible.

In the case of medicinal and aromatic plants, 17 surveys were performed in Santos Lugares, 30 surveys in Guampacha and 80 surveys on a sample population within the 18-82 age range, being women the majority. [16, 15]

Regarding dye plants, the sample population to which the survey was conducted was composed of thirty-five (35) people in the 16-78 age range, and corresponding entirely to women (no man was found dedicated to the craft with sheep wool or to the collection of dye plants) [19].

METHOD

The Contingent Valuation Method (CVM) was applied in this research. This method is a stated preference method implemented by means of surveys, and aims at assessing how individuals would hypothetically react to changes in environmental quality. In particular, it finds out how much respondents would be willing to pay for improved environmental quality or to avoid a hypothetical reduction in environmental quality [23].

The welfare of an individual (U) depends on the real income to buy goods and services, which is represented in the following expression [24, 25]:

$$\text{or } U_1 = b(Yr) + e_0 \quad \textbf{(1)}$$

where Yr is the real income of the individual and b is the marginal utility that depends on real income. According to Díaz [15], this equation explains that an increase in the individual's income increases their comfort level and a decrease in income causes a decrease of the same. Taking into account the environmental goods and services that do not have a defined market, the recognition of its importance in individual welfare changes the equation as follows:

$$U = Ba(s) + b(Yr) + e_0 \quad \textbf{(2)}$$

where Ba is the welfare of individuals for the enjoyment (or not) of the environment, which depends on the socioeconomic and cultural characteristics of the people (represented by s), and e0 is a random variable with mean zero referring to taste different from the average of the area studied. If the current quality of a resource and environmental services is to be improved or maintained for the individuals, a payment (P) is proposed for their conservation. In this case the utility function would now be as follows:

$$U' = B'a(s) + b(Yr - P) + e_1 \quad \textbf{(3)}$$

This means that in this case the marginal utility will arise from the difference between actual income (Yr) and the amount that the individual is willing or is required to pay (P) to carry out conservation. That is why for the individuals to experience an increase in their welfare caused by conserving their resources, their welfare after payment must be greater than the initial situation:

$$U' - U = B'a(s) - Ba(s) + b(Yr - P) - b(Yr) + e_1 - e_0 > 0 \quad \textbf{(4)}$$

According to Rojas Padilla *et al.* [24] and Diaz [15], there is an amount of money that users are willing to pay that makes the U' and U functions equal, but with a higher level of well-being. They also note that studies on WTP must calculate such amount and the factors that determine it. Finally, the role of valuation of assets is estimated, and this is done by a simple regression whose equation is:

$$WTP = a + bX + e \quad \textbf{(5)}$$

where a and b are the set of parameters to be estimated, X is the matrix of socioeconomic characteristics of individuals and e is the error term.

Application

127 families of six locations were surveyed to investigate the use of medicinal and aromatic plants and 35 surveys were conducted to assess the use of dye plants for wool weave. The surveys were performed between 2005 and 2008.

The variables willingness to pay (WTP) and willingness to accept (WTA) were used in this study. The data were analyzed using SPSS. A demand curve was constructed and Consumer Surplus for willingness to pay data was calculated.

The value of the individual consumer surplus was obtained from the value of the consumer surplus of the whole sample. Then, the individual consumer surplus was extrapolated to the population to calculate the population consumer surplus and, finally, the annual consumer surplus for 12 months.

RESULTS AND DISCUSSION

This section presents the results obtained from the data collected and analyzed in 4 locations (Villa Figueroa, El Empachao, Blanca Pozo and San Jerónimo). Only the results of these towns are presented for space reasons.

The results show the values of Willingness to Pay (WTP), maximum willingness to pay, and the descriptive statistics of the variables. The calculation of the Consumer Surplus (CS) and the three functions related to mathematical models are also presented here. Fig. (**2**) shows the values expressed by 80 respondents in 4 different locations of Santiago del Estero [22]. The final values of all studies were converted into US currency exchange rate at that time: US $1 = $4,366 [25].

About 40% of the respondents would be willing to pay 5 pesos (US $1.14) to preserve the plants while 16.3% expressed willingness to pay 20 pesos (US $4.58).

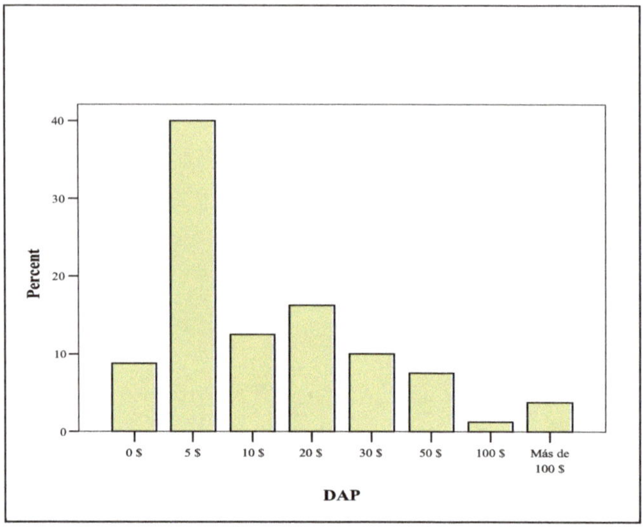

Fig. (2). Values of willingness to pay (DAP in Spanish) in 4 locations in Santiago del Estero (expressed in Argentine pesos).

Table 1. Descriptive statistics of MaxWTP (MaxDAP).

	N	Minimum	Maximum	Mean	Std. Deviation
MAXDAP	80	$0	$120	$18.69	$24.219
Valid N (listwise)	80				

After the descriptive statistics analysis, an analysis of multiple linear regression using the "stepwise" method was performed to determine which variables influence the value on the maximum WTP. With the SPSS statistical program, two models have been obtained with predictor variables and error data indicating that the variables that influence the MaxWTP are: the level of household income, and income and age of the respondents together (Table **2**).

Table 2. Influential Variables in MaxWTP (MaxDAP).

Model Summary

Model	R	R Square	Adjusted R Square	Std. Error of the Estimate
1	,310[a]	,096	,085	$23.169
2	,393[b]	,154	,132	$22.562

a. Predictors: (Constant), Ingresos

b. Predictors: (Constant), Ingresos, Edad

The models show high levels of significance and confirm the impact of family income *(ingresos)* and age *(edad)* on the other hand. The positive sign of the income variable indicates that the higher the household income, the higher the MaxWTP. The negative sign of the variable age indicates that the higher this is, the lower the MaxWTP. The remaining variables were excluded from the study since they were not significant. They are presented in Table **3**.

Table 3. Multiple Linear Regression on Variable MaxWTP.

Coefficients[a]

Model		Unstandardized Coefficients		Standardized Coefficients	t	Sig.
		B	Std. Error	Beta		
1	(Constant)	-5,771	8,867		-,651	,517
	income	5,420	1,879	,310	2,884	,005
2	(Constant)	7,837	10,479		,748	,457
	income	5,955	1,845	,341	3,228	,002
	age	-,350	,153	-,242	-2,292	,025

a. Dependent Variable: MAXDAP

Table 4. Linear regression on socio-economic variables.

Excluded Variables[c]

Model		Beta In	t	Sig.	Partial Correlation	Collinearity Statistics
						Tolerance
1	sexo	-,104[a]	-,962	,339	-,109	,986
	edad	-,242[a]	-2,292	,025	-,253	,984
	estado.c	,003[a]	,031	,975	,004	,998
	nºfamili	-,080[a]	-,737	,463	-,084	,980
	años.res	-,117[a]	-1,084	,282	-,123	,999
	tenencia	-,087[a]	-,792	,431	-,090	,959
	n.educac	-,006[a]	-,052	,959	-,006	,982
2	sexo	-,131[b]	-1,242	,218	-,141	,975
	estado.c	,007[b]	,065	,948	,007	,997
	nºfamili	-,107[b]	-1,001	,320	-,114	,969
	años.res	,168[b]	1,023	,309	,117	,407
	tenencia	-,080[b]	-,743	,460	-,085	,958
	n.educac	,047[b]	,431	,667	,049	,939

a. Predictors in the Model: (Constant), Ingresos

b. Predictors in the Model: (Constant), Ingresos, Edad

c. Dependent Variable: MAXDAP

Next, a multiple linear regression analysis (Table **4**) with socioeconomic variables (gender, age, marital status, number of family members, years of residence in the location, land tenure situation, educational level) was performed to determine which variables affected the WTP most.

Once the influential variables were identified, the models were constructed using the coefficients of the second column in Table **2** for:

MODEL 1: *AXWTPVALUE= -5,771 + 5,420 * Income*

MODEL 2: *AXWTPVALUE = 7,837 + 5,955 * Income – 0,350 * Age*

The next step was to find the value of consumer surplus (CS). The CS is determined by the area under the demand curve. This calculation allows expressing what the economic value of the use of medicinal plants is for the studied community. To do this we proceeded to find, first, the demand curve. The functions with their corresponding R^2 are as follows (Table **5**).

Table 5. Comparison between models, function and R^2.

MODEL	FUNCTION	R^2
Linear	$y = -0.8005x + 51.106$	0.5899
Logarithmically	$y = -25.24\ln(x) + 105.03$	0.9135
Polynomial	$y = 0.0241x^2 - 2.7504x + 77.755$	0.8174

The equation of the demand curve is: $y = -25.24\ \ln(x) + 105.03$

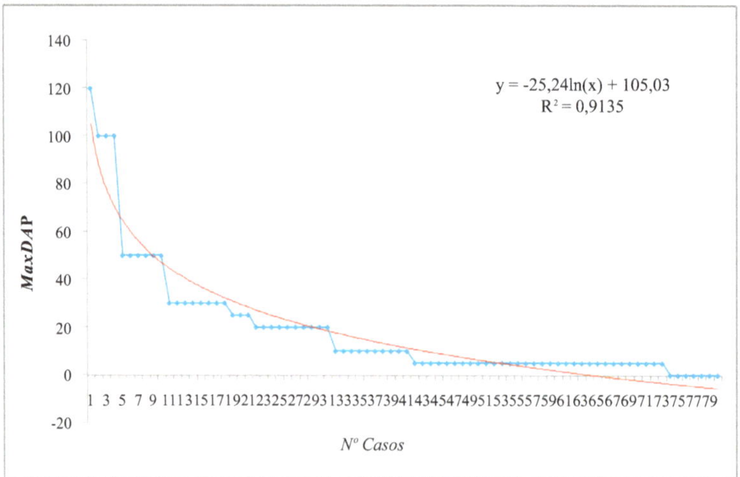

Fig. (3). Demand curve with values of willingness to pay by respondents.

The result of the equation represents the area under the curve which, in turn, expresses the CS. This value (in Argentine pesos) is obtained by applying the integral function for:

$$\int_{0}^{80} -25.24\ln(x)+105.03dx \quad \textbf{(7)}$$

CS= $ 1573,41

The next step was to find the value of Individual Consumer Surplus (ECI), which is obtained through the following equation:

$$ICS = CS/n \quad \textbf{(8)}$$

where *n* is the sample size. It must be reminded that in this case the sample size is 80, corresponding to the valid number of surveys, thus obtaining

$$ICS = 1573{,}41/80 = \$\ 19{,}67 \quad \textbf{(9)}$$

Then the extrapolation to the whole population is performed to calculate the Total Consumer Surplus (TCS) [26].

$$TCS = ICS * N \quad \textbf{(10)}$$

where N is the size of the population. It must be remembered that the total population of the study areas is 111 families, so that: *TCS = 19.67* 111 = $2183.37.*

Finally, Consumer Surplus Annual Total (TACS) is calculated: *TACS = TCS * 12*

So, the following value is obtained: *ECAT = 2183.37 * 12 = 26200.44*

This final value, expressed in Argentine pesos, will be the amount of money that the population of the three sampled areas would be willing to pay annually for the medicinal use of flora. In other words, the benefits are expressed in monetary units; this amount represents the benefits that people obtain from the use and conservation of medicinal plants (Table **6**).

Table 6. Comparison of results by location, year, variable, mean and total value.

Place and year	Sample size	Population size	Method	Variable	Mean value	Total Value Dollar/year (Argentine peso)
Medicinal and aromatic plants						
Santos Lugares 2005 [16]	17	200	Contingent Valuation/ Avoided Cost	WTP Avoided cost	**US $34.35/month** ($150.03/month)	**US $82,471.82/year** ($360,072/year)
Guampacha 2007 [15]	30	200	Contingent Valuation	WTP	**US $32.59/month** ($142.33/month)	**US $78,239.48/year** ($341,593.6/year)
Villa Figueroa 2008 [19, 22]	80	39	Contingent Valuation	WTP	**US $4.50/month** ($19.67/month)	**US $6,000.91/year** ($26,200.14/year)
El Empachao 2008 [19, 22]		48	Contingent Valuation			
San Jerónimo 2008 [19, 22]		24	Contingent Valuation	WTA	**US $5.07/month** ($22.15/month)	**US $6,760.83/year** ($29511.79/year)
Blanca Pozo 2008 [19, 22]			Contingent Valuation			
Dye plants						
San Jerónimo 2008 [19, 22]	35	35	Contingent Valuation	WTP	**US $133.44/month** ($582.63/month)	**US $448,382.22/year** ($1,957,636.8/year)
Blanca Pozo 2008 [19, 22]			Contingent Valuation	WTA	**US $4.18/month** ($18.27/month)	**US $14,066.87/year** ($61,415.96/year)

In most cases the plants obtained come firstly from the local and/or native vegetation and secondly from individual crops or gardens. Many plant species have a potential for multiple uses. However, little research has been done on the identification and utilization of non-timber tree parts. Research on the utilization of plant parts will help to identify new uses and improve the production of already known products [27].

CONCLUDING REMARKS

Regarding medicinal species, there are only two other studies in the province and correspond to the western area departments. This work was the first carried out in the sampled areas and the results can be compared with the ones obtained in other studies and will help expand the database in Santiago del Estero.

Regarding plants for medicinal purposes, of the total respondents, 97.5% obtained them directly from the forest. In addition, 65% said to have cultivated a plant in their garden. Most people obtain the plants from the forest or grow them in their garden. The importance of the natural environment for these people as a source of alternative timber resources is demonstrated. The results can be the basis for researching the development of potential markets of medicinal plants in the region, as Khan and Abdur comment [28].

More research on the use of aromatic plants in Santiago del Estero should be done since there are no research results related to the valuation of aromatic plants in the province.

Regarding dye species, there is no history of research in the province. This document is the first of its kind in this area. The importance of the implementation of the method and the value found can generate future applications in similar situations. In Salta, Argentina, Lambaré *et al.* [29] studied and compared the use of dye plants between 1994 and 2000 with those in 2007 and 2008. They also analyzed the current factors in the acquisition and transmission of knowledge related to such use.

In relation to rural people's perceptions of the value of natural resources, the healing uses they give to medicinal plants have become a natural heritage they are willing to protect and preserve for future generations.

As for the economic value expressed in monetary terms, the environmental service of the plants studied represents money that people avoid spending on medical services due to the free availability of medicinal plants.

By comparing the average value MaxWTP already shown in Table **1** ($18.69) with ICS ($19.67), an analogy between these values is observed, which confirms the accuracy of the statistical analysis.

If the values of Consumer Surplus and Surplus Compensatory calculated for the WTP and WTC are contrasted, it can be noted that the values obtained are very similar for both, although slightly higher than those calculated for WTA and WTP.

CONFLICT OF INTEREST

The authors declare no conflict of interest, financial or otherwise.

ACKNOWLEDGEMENTS

Declared none.

REFERENCES

[1] FAO. "Productos Forestales no Madereros; posibilidades futuras" Estudio FAO Montes 97. Roma: Organizaciòn de las Naciones Unidas para la Agricultura y la Alimentación 1992.

[2] El Mokni R, Mahmoudi MR, El Aouni MH. Contribution a la valorization de certaines plantes aromatiques et medicinales infoedeesauxmonts de la Kroumirie, Nord-ouest de la Tunisie. Acta Hortic 2013; (997): 245-50. [http://dx.doi.org/10.17660/ActaHortic.2013.997.31]

[3] Boyd J, Banzhaf S. What are ecosystem services? The need for standardized environmental accounting units. Ecol Econ 2007; 63: 616-26. [http://dx.doi.org/10.1016/j.ecolecon.2007.01.002]

[4] Cadena-González AL, Sørensen M, Theilade I. Use and valuation of native and introduced medicinal plant species in Campo Hermoso and Zetaquira, Boyacá, Colombia. J Ethnobiol Ethnomed 2013; 9: 23. [http://dx.doi.org/10.1186/1746-4269-9-23] [PMID: 23578098]

[5] Venn T, Quiggin J. Accommodating indigenous cultural heritage values in resource assessment: Cape York Peninsula and the Murray–Darling Basin, Australia. Ecological Economics 2007; 61: 334-44. [http://dx.doi.org/10.1016/j.ecolecon.2006.03.003]

[6] Allen K, Moore R. Moving beyond the exchange value in the nonmarket valuation of ecosystem services. Ecosyst Serv 2016; 18: 78-86. [http://dx.doi.org/10.1016/j.ecoser.2016.02.002]

[7] Stenger A, Harou P, Navrud S. Valuing environmental goods and services derived from the forests. J For Econ 2009; 15: 1-14. [http://dx.doi.org/10.1016/j.jfe.2008.03.001]

[8] Winkler R. Valuation of ecosystem goods and services. Part 1: An integrated dynamic approach. Ecol Econ 2006; 59: 82-93. [http://dx.doi.org/10.1016/j.ecolecon.2005.10.003]

[9] Sarmiento M. Desarrollo de un nuevo método de valoración medioambiental. PhD Thesis Universidad Politécnica de Madrid España 2003.

[10] Tinitana F, Ríos M, Romero-Benavides JC, de la Cruz Rot M, Pardo-de-Santayana M. Medicinal plants sold at traditional markets in southern Ecuador. J Ethnobiol Ethnomed 2016; 12(1): 29. [http://dx.doi.org/10.1186/s13002-016-0100-4] [PMID: 27380631]

[11] Pert PL, Ens EJ, Locke J, Clarke PA, Packer JM, Turpin G. An online spatial database of Australian Indigenous Biocultural Knowledge for contemporary natural and cultural resource management. Sci Total Environ 2015; 534: 110-21. [http://dx.doi.org/10.1016/j.scitotenv.2015.01.073] [PMID: 25682266]

[12] Barata A, Rocha F, Lopes V, Carvalho A. Conservation and sustainable uses of medicinal and aromatic plants genetic resources on the worldwide for human welfare. Ind Crops Prod 2016; 88: 8-11. [http://dx.doi.org/10.1016/j.indcrop.2016.02.035]

[13] Morales S, Ladio A. Medicinal plants in a Mapuche community from NW Patagonia Argentina: Classification and organoleptic perceptions related to its valuation. Bol Latinoam Caribe Plantas Med Aromat 2008; 7(3): 149-55.

[14] Morales S, Ladio A. Edible and medicinal plant complexes in Patagonia Argentina: Components and possible associated processes. Bol Latinoam Caribe Plantas Med Aromat 2015; 14(3): 237-50.

[15] Díaz R. Valoración económica de servicios ambientales: uso medicinal de la vegetación en las Sierras de Guasayán Santiago del Estero, Argentina Graduation Thesis Universidad Nacional de Santiago del Estero Argentina 2007.

[16] Sarmiento M, Borselino S, González A. Aproximación al valor económico ambiental de las

propiedades medicinales de plantas nativas en el Chaco Semiárido III Congreso Forestal Argentino y Latinoamericano 2005 Corrientes Argentina.

[17] Sarmiento M, Barraza J, Bonacina E, Cuello S, Escalada C, Marelli S. Conocimiento de productos alimentarios del monte nativo por la población urbana en Santiago del Estero, Argentina. Resumen científico V Jornadas Forestales de Santiago del Estero y I Feria Foresto-Industrial Santiago del Estero 2015.

[18] Sarmiento M, Mazur M, Neme A, Pintos S, Romano L, Ruiz P. Preferencias declaradas en el consumo de bienes naturales y productos derivados del monte nativo por pobladores urbanos en Santiago del Estero. Resumen científico V Jornadas Forestales de Santiago del Estero y I Feria Foresto-Industrial Santiago del Estero 2015.

[19] Sarmiento M, Guerrero Maldonado N. Valoración económica de servicios ambientales proporcionados por plantas tintóreas en bosques nativos de la región chaqueña en Santiago del Estero Argentina XIII Congreso Forestal Mundial 2009 Buenos Aires Argentina.

[20] Arraiza MP, Brutti O, Farías G, *et al.* Creación de una Red Temática de Plantas Medicinales Aromáticas y Tintóreas (RT_PMAT) en el Norte Argentino. Quebracho (Santiago del Estero) 2008; 71-73(15)

[21] INDEC. Censo Nacional de Población y Viviendas 2010. Argentina: Instituto Nacional de Estadísticas y Censos 2010.

[22] Guerrero Maldonado N. Uso y Valoración de Plantas Medicinales y Tintóreas presentes en Santiago del Estero, Argentina. Proyecto de fin de carrera Escuela Técnica Superior de Ingenieros de Montes Universidad Politécnica de Madrid España 2008.

[23] Luangmany D, Voravong S, Thanthathep K, Souphonphacdy D, Baylatry M. Valuing Environmental Services Using Contingent Valuation Method. Economy and Environment Program for Southeast Asia 2009. Available at: http://econpapers.repec.org/paper/eeptpaper/tp200910t4.htm

[24] Rojas Padilla J, Pérez Rincón M, Peña Varón M. "La valoración contingente: una alternativa para determinar la viabilidad financiera de proyecto de tratamientos de aguas residuales en zonas rurales de países tropicales" Info Andina Consorcio para el Desarrollo Sostenible de la Ecorregión Andina. Cali, Colombia: CONDESAN 2001.

[25] FIDE. Evolución de la Cotización del Dólar Oficial Valores constantes 2012.

[26] Sarmiento M. Desarrollo de un nuevo método de valoración medioambiental. Tesis Doctoral Universidad Politécnica de Madrid 2003.

[27] Sarmiento M, Guerrero N, Díaz R. Valoración y propuesta de gestión de servicios ambientales de plantas medicinales y tintóreas en comunidades rurales de Santiago del Estero, Argentina. Primer Simposio Internacional El monitoreo de Servicios Eco-sistémicos en los mecanismos de compensación por pagos por Servicios Ambientales Ciudad de Guatemala 2008.

[28] Khan H, Abdur R. Medicinal Plants: Economic Perspective and Recent Developments. World Appl Sci J 2014; 31(11): 1925-9. [http://dx.doi.org/10.5829/idosi.wasj.2014.31.11.14494]

[29] Lambaré D, Hilgert N, Ramos R. Dyeing Plants and Knowledge Transfer in the Yungas Communities of Northwest Argentina. Econ Bot 2011; 65: 315. [http://dx.doi.org/10.1007/s12231-011-9169-9]

CHAPTER 10

Principles of Medicinal and Aromatic Plant (MAPs) Propagation, Production and Collection; *Artemisia absinthium* L.

Jesús Burillo[1,*], **Azucena González-Coloma**[2], **Javier Tapia**[3] and **Juliana Navarro Rocha**[1]

[1] *Departamento de Ciencia, Tecnología y Universidad, Centro de Investigación y Tecnología Agroalimentaria de Aragón, Gobierno de Aragón, Avda. Montañana, 930, Zaragoza, Spain*

[2] *Instituto de Productos Naturales y Agrobiología, CSIC. Avda. Astrofísico F. Sánchez 2, 38206 La Laguna, Tenerife, Spain*

[3] *Departamento de Económico, Facultad de Economía y Empresa, Universidad de Zaragoza. Gran vía 2, Zaragoza, Spain*

Abstract: Wormwood *(Artemisia absinthium* L.) is a perennial plant known for its antifungal, parasitic and nematicidal effects. Its essential oil is rich in thujone which limits its industrial use due to neurotoxic effects. With the aim of domesticating plants free of thujone without sacrificing essential oil levels for use as a potential biopesticide, the study was conducted during 8 growing seasons at an experimental field located in Ejea de los Caballeros, Aragón, Spain. It compared two populations: the original population of wild plants collected in the Sierra Nevada National Park (Gr-75) and the selected population from a pre-trial in Teruel (Te-68). The trial was arranged according to a randomized block design with 3 replications. The experimental plot consisted of 6 rows of plants 20 meters long with 738 plants per population. Plants were harvested every year except for year one. The following variables were evaluated: production (in kilograms of fresh and dry biomass), essential oil obtained from steam-distilled biomass (l/ha) and plant mortality. All variables were checked annually. According to yearly averages, the Teruel population maintained a stable percentage of dry material with higher production of essential oil compared to the Granada population (39 l/ha on average). It was also apparent that essential oil production has a significant direct correlation with dry material. The result of the study is the ®Candial variety registered under Decision No EU 36714 of 27 January 2014 at the Community Plant Varieties Office.

Keywords: *Artemisia absinthium L.*, Domestication, Experimentation, Production and variety.

* **Corresponding author Jesús Burillo:** Departamento de Ciencia, Tecnología y Universidad, Centro de Investigación y Tecnología Agroalimentaria de Aragón, Gobierno de Aragón, Avda. Montañana, 930, Zaragoza, Spain; Tel: +34 976716374; E-mail: jburilloa@aragon.es

M. Paz Arraiza, A. González-Coloma, J. Burillo and C. Calderón-Guerrero (Eds.)

INTRODUCTION

Cultivation of aromatic and medicinal plants is particularly important in the Community of Aragon in light of the abundance of wild aromatic-medicinal flora [1]. The selection of chemotypes able to adapt to these soil and climatic conditions along with suitable cultivation techniques can provide an alternative to dominant local crops, *i.e.* cereals. These studies take on even greater importance considering the potential of these plants to produce products of industrial interest, especially since Spain and the European Community as a whole lack such raw materials.

Wormwood or absinthe *(Artemisia absinthium* L.) is a species known and used since ancient times. Galen, Hippocrates and others described many of the medicinal properties of absinthe and the Dioscorides treatise on medicinal substances compiles and expands upon the medicinal qualities of this plant [2] and use of the plant's extracts to produce an alcoholic drink which was very popular in the 19th century.

Today this drink is prohibited due to the neurotoxic effect of α-thujone present in many essential oils of *A. absinthium* [1]. The pharmaceutical industry uses the plant in preparations administered orally and topically [3].

It is a perennial, woody strain with herbaceous branches that can reach a height of over one meter. The leaves are gray-white on the underside and light green on top. Flowers are tubular and pale yellow in color. The bracts are light fresh. It can be propagated by cuttings or seeds with an average germination rate of 80% in the laboratory. Best results are achieved by planting nursery seedlings. Leaves and flowering tops are the parts used [1].

This species is resistant to winter frost and is productive as of year one after planting. It also fully adapts to mechanized farming techniques. In Spain it forms part of the native flora in mountain areas as a ruderal species. In Aragon it is most abundant at an altitude of 600-700 m in the Pyrenees foothills, Jiloca Valley and the district of Teruel where it grows on uncultivated land or embankments [1].

This study on the perennial Wormwood plant *(A. absinthium)* forms part of the experimental scheme focusing on aromatic and medicinal plants in Aragon [4]. One of the goals of this study is to gain greater insight into its agronomic characteristics and to develop a standard plant [5] to determine its likelihood of serving as a natural insecticide, and especially to develop a thujone-free variety making it more readily acceptable for use in the food industry.

Cultivation of this species requires the selection of plants adapted to the area and that are also chemically stable. This entails a systematic process of domestication.

Domestication is a process of acclimating plant species to specific man-made growth conditions. During this domestication process, wild plants are initially subjected to some sort of management, like a study of drought and/or type of soil in pots. Plants are then sown in the field. At this final stage, the process involves selection and improvement of genotypes resulting in a population of relatively uniform plants with a more restricted genetic base. Simply stated, the domestication process entails changing the biological characteristics of the species or, in a broader sense, it can be described as a change in the way the species is used which brings about morphological and genetic changes in the plants [6, 7].

Domestication implies improving the quality of the raw material allowing for standardization of its active ingredients and homogeneous plant populations required for production, processing and use. Moreover, we use the term 'introduction of plants' which, depending on how the plant responds to improvements, means cultivating a domesticated species in a new location which is some distance from the geographical area where the plants were originally found. However, this same term can be used to refer to the cultivation of non-domesticated species [8].

The specific objectives of this study are: to evaluate the plant's adaptation to the trial area, gain greater insight into production potential and economic viability and to study yield and quality of dry material and essential oils from a process of drying and distillation.

MATERIAL AND METHODS

Geographical Considerations

The domestication process of wormwood *(A. absinthium)* entailed finding a population with a low percentage (traces) of α and β-thujones, main components of its essential oil [9 - 12]. Taking soil and climate variability into consideration, the process began in different experimental plots in the Autonomous Community of Aragon in 2001. The results of these preliminary-trials set the stage for the beginning of the domestication process planting pre-selected master stock in a plot in Teruel (Aragon, Spain). The experiment itself began in 2008 and continued until 2014 (seven-year trial) at Ejea de los Caballeros, Aragón, Spain.

The coordinates of the plot at Ejea de los Caballeros are 42°8'8.73" N, 1°12'31.50" W, at an altitude of 346 m.a.s.l. The soil has a clay-loam texture as described in Table **1**, determined through a series of soil analyses conducted before and every two years during the trial. Organic fertilizer (BIOHUMI®, with 1.98% of organic Nitrogen) was applied during the intermediate years (1,500 Kg/ha).

Table 1. Physical and chemical properties of soil - experimental area Ejea de los Caballeros, Zaragoza.

Property	Units	Depth	
		30cm	50cm
Sand Silt Clay	%	30.82 31.10 37.97	38.88 30.30 30.81
Water pH EC	dS/m	8.7 ± 0.3 0.2 ± 0.03	8.8 ± 0.3 0.2 ± 0.03
O.M.	g/100g	2.91 ± 0.37	2.13 ± 0.27
P (soluble) -Olsen K Mg	Mg/kg	85 ± 17 490.0 ± 75.07 206.0 ± 20.19	46 ± 9 242.0 ± 37.07 198.0 ± 19.4

EC: electrical conductivity; O.M.: organic matter; P: phosphorus soluble in sodium bicarbonate (spectrophotometer); K: potassium; Mg: magnesium.

Given the low rainfall in the area, drip irrigation was used and extra water was occasionally provided during the summer months characterized by high evapotraspirative demand and scant precipitation. Drip irrigation was set to 4 l/hour for 5-6 hours per week. Spacing between drippers was 30 cm.

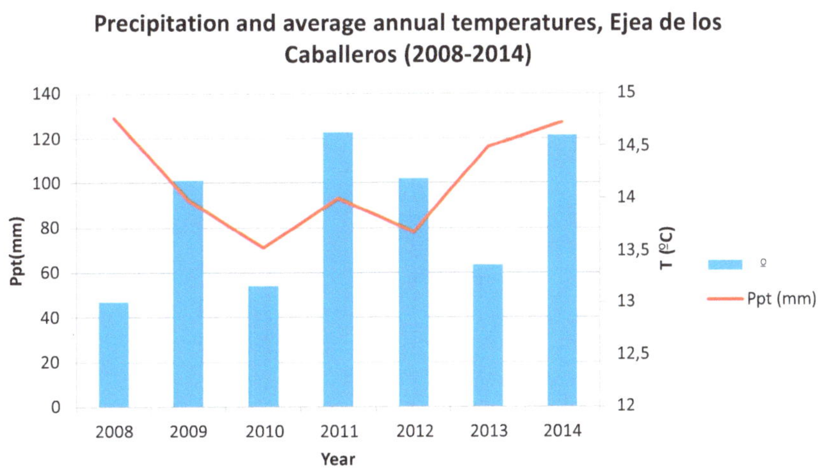

Diag. (1). Summary of weather conditions at Ejea de los Caballeros (2008-2014), annual averages.

Diag. (**1**) shows weather conditions at Ejea de los Caballeros, during the various growing seasons (2008-2014).

Technical Considerations

The trial crop was planted on 26 and 27 April 2007 and arranged according to a randomized block design with 3 replications and with two references populations, the original population of wild plants collected in the Sierra Nevada National Park (Gr-75) and the population created from pre-selected material from a trial previously conducted in Teruel (Te-68). Plot size was 1.5 m x 0.5 m, *i.e.* 0.75 m^2 per plant or 13,333 plants per hectare with no replanting as of year two. The experimental plot consisted of 6 rows of plants 20 meters long with 738 plants per population. Fig. (**1**) is a planting sketch and Figs. (**2**, **3** and **4**) represent different moments during the trial.

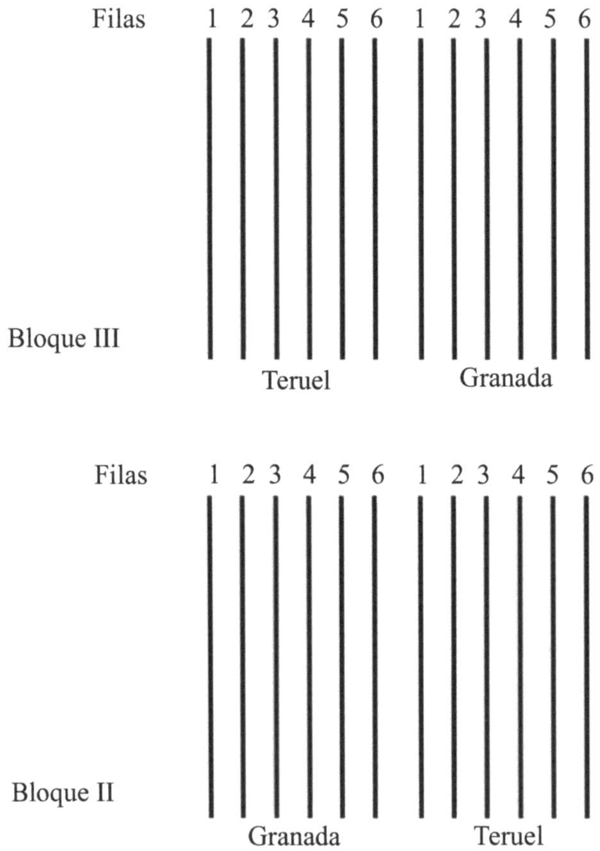

Fig. 1 contd.....

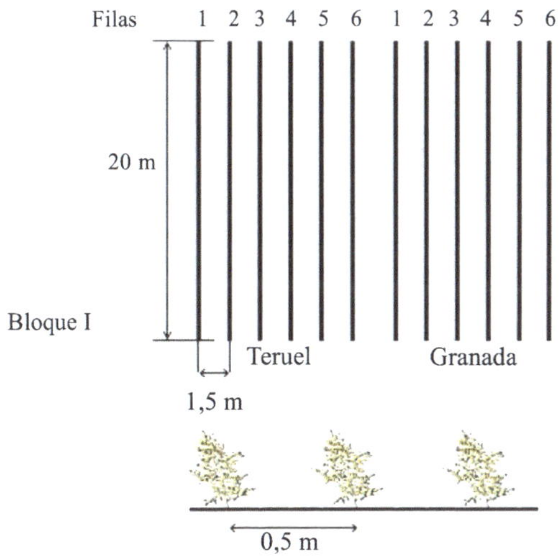

Fig. (1). Sketch of the experimental plot.

The experimental crop was grown for a total of eight years. The crop was planted in 2007 and was maintained until 2014, the last year it was harvested. Plants were harvested every year except for year one.

The following variables were considered for each type of population: crop production and essential oil obtained from the distillation of biomass. All variables were checked annually.

The crop production variable refers to the weight of fresh biomass harvested from wormwood plants annually in August or September. Production of fresh biomass is measured by means of outer sampling, inner sampling and the total harvest. For outer sampling, the first five plants of the third and fourth rows in each block were selected. For inner sampling, the next five plants of the same rows were selected. Total harvest data encompasses both outer and inner sampling and the total from the rest of the wormwood plantation.

The production of fresh biomass is measured in kilograms per plant and kilograms per hectare. For each study group, outer and inner sampling or total harvest is used to calculate production which is then divided by the number of plants to calculate the production (in kilograms) of each plant. According to the planting scheme used in the study, calculation of per-hectare production is based on 13,333 wormwood plants per hectare. This is the figure used to calculate total theoretical production per hectare.

Fig. (2). Wormwood cycle at the Ejea experimental plot.

Fig. (3). Close-up of flowering wormwood plants.

Fig. (4). Detail of flowering tops of the wormwood plant.

Steam distillation was used to extract essential oil from the fresh biomass harvested from wormwood plants. We used a low and high pressure pilot extraction plant available to us at the Center for Food Research and Technology (CITA) in Zaragoza (Aragón, Spain), being 74 km from the field. The plant can fractionate the different extractions at different pressures and temperatures in order to determine both the yield and quality of the volatile compounds obtained. The pilot plant is also equipped with an ethanolic extraction module for heavier compounds not obtained by steam distillation and which are of great interest to industry. Steam distillation entails producing steam in a generator that circulates it through an extraction vessel under controlled pressure and temperature. The steam draws out all the volatile compounds contained in the biomass which arrive to the coil in the form of water vapor where they are cooled and condensed and are subsequently collected in a decanter which separates the water from the essential oil by density difference (Figs. **5-6**).

Fig. (5). General view of the pilot extraction plant used.

Fig. (6). Detail of the pilot plant with its two extraction modules.

To control quality, distillations are repeated in small quantities in the laboratory by means of hydrodistillation using a Clevenger trap [13] included in the European Pharmacopoeia (Fig. 7).

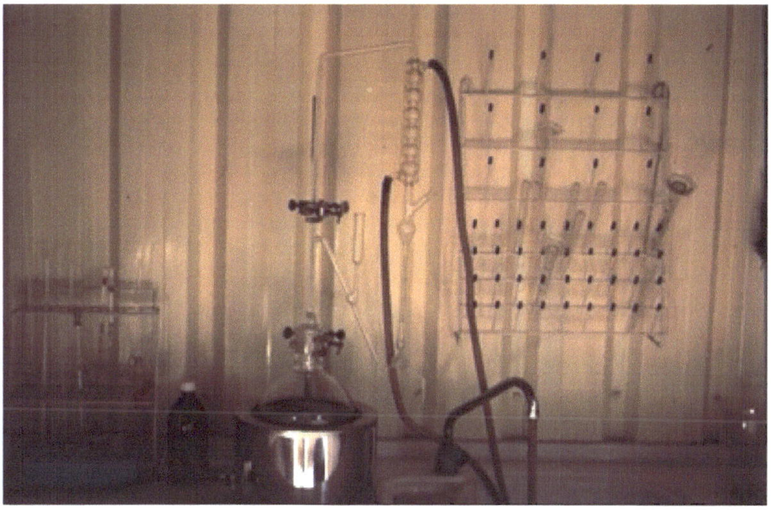

Fig. (7). Detail of Clevenger hydrodistillation.

As only a small amount of material was collected in the samplings, we ruled out the option of distilling sampling biomass separately, except in 2009 for the Te-68 population. Therefore, we only have specific population essential oil values for that one year and population. Essential oil production is measured in liters per plant and in liters per hectare and all essential oil extracted from total production each year is quantified.

A univariate analysis was conducted for each variable, drawing a distinction between the two populations, Granada and Teruel, where position, dispersion and shape were examined [14]. However, the representativeness of the study's measurements and the values themselves must be interpreted with caution due to the low number of observations (seven in total) and the experimental nature of the wormwood plantation [15].

A sample from inner rows 3 and 4 was selected each year. Samples were taken from the first five plants of each of those rows, plants one to five (outer sample) and from the next five plants in each row, plants six to ten (inner sample). Harvest data includes all of the rest of the plants plus the sample data collected.

Anova test were made for the most important variables biomass and oil production. In both cases, yearly productions per hectare of the variables were considered for the study.

RESULTS

General Data and Statistical Analysis

a. Granada population (Tables **2-5**).

Fresh Biomass

The production of fresh biomass sampled per plant varies between 1.6 and 2.63 kg, the average during productive years being 1.86 kg. The quasi-standard sampling deviation is 0.358.

The average harvest is outside the range obtained from sampling. However, when calculating an estimate in the form of a confidence interval for the harvest mean with a 95% confidence interval the range is between 1,532 and 2,194 such that the harvest mean of 1,534 is within the range thus indicating that sample production does provide a good estimate of harvest production.

The value obtained for the second year of production, *i.e.* 2.63 kg in 2009, is clearly atypical as it lies outside the confidence interval introducing a degree of positive asymmetry in the distribution.

Dry Material

The amount of dry material harvested was not measured independently. Instead, we compared the production of fresh and dry material sampled each year and came up with yearly average percentages of dry biomass as a product of fresh biomass. The annual amount of dry material harvested is then extrapolated from these percentages.

In drawing a distinction between per-plant and per-hectare production (both in kilograms), the range of values of analyzed data is very small, 0.33 kg/plant and 4,424 kg/hectare, both around a half of the minimum values of series, which it indicates high concentration of data.

Essential Oils

Essential oil production was calculated only for the total harvest due to the low volume of sample biomass collected. The range was 1.04 ml/plant with a minimum yield of 1.35 and maximum of 2.40 ml/plant. An average 2.02 ml of essential oil per plant was obtained which translates into a mean per-hectare yield of 29.80 liters, the standard deviation being 0.37 ml/plant and 6.43 l/hectare.

Pearson's variation coefficient gives a value of 0.158 if it is being considering output per plant and 0.188 considering output per hectare. Then mean can be considered representative of the output's production along the crop's lifetime. In principle, Pearson's variation coefficient does not vary if there is a change of scale and therefore distribution measured in terms of plants or hectares should theoretically provide the same result. However, due to plant mortality throughout the study, per-plant production is not obtained by dividing total production by the same number of plants each year. Therefore, the shift between per-plant and per-hectare production cannot be considered simply as a change of scale.

The median yield is 2.09 ml/plant and 29.78 l/ha. The mean and median are close, indicating that the population is clustered around the mean value without extreme values.

Lastly, production figures from year two (1.47 ml/plant and 18.06 l/ha) are surprisingly low. Despite an increase in the production of fresh material that year, essential oil yield was lower which would appear to indicate a negative correlation between these two variables once the plant has reached a certain stage in its development.

Essential oil figures per production have not been discussed because they are not comparable between different years due to plant mortality in recent years. In other words, figures from different years do not correspond to the same number of plants. Comparisons can be drawn by dividing production by the number of plants actually contributing to that production (production per plant) as analyzed above.

Plant Mortality

Monitoring of plant mortality was important throughout the experiment with both populations due to the high mortality rate in both populations in 2012 and 2013. It is important to remember that no plants were replaced during the course of the experiment, not even in the early years. In any case, plant mortality was highest in year five and six (production years four and five) and therefore it would have made little sense to replace dead plants with new ones which would not begin to produce until the following year and would skew all of the measurements taken per plant. This is why the number of plants in the 2013 and 2014 production figures is lower, for both the whole population and the sample as some of the plants lost were from sampling rows 3 and 4. Mean figures were calculated based on fewer individuals during those years and could account for differences in previous years' production.

Table **5** shows plant mortality in number of individuals by block and row. The two populations behaved in a similar fashion, *i.e.* zero mortality until 2011 when a fair number of plants died. The number of plants lost increased four-fold the following year and that number doubled again in 2013 (31 plants in the case of the Granada population). No more losses were registered after 2013 and calculation of the different variables was based on the number of remaining plants so as to maintain year-to-year comparability.

Table 2. Granada population General data.

	2008	2009	2010	2011	2012	2013	2014	mean
Production per plant (l)	1.36	1.55	1.40	1.74	1.31	1.78	1.61	1.53
Production per hectare (l/ha)	18,104	20,622	18,717	23,141	17,402	23,764	21,456	20,458
Essential oil per plant (l)	$2.31x10^{-3}$	$1.36x10^{-3}$	$1.60x10^{-3}$	$2.09x10^{-3}$	$2.11x10^{-3}$	$2.40x10^{-3}$	$2.35\ x10^{-3}$	$2.03\ x10^{-3}$
Essential oil per hectare	30.77	18.07	21.34	27.82	37.41	31.98	34.20	28.80
Essential oil per harvest (l/ha)	0.170	0.102	0.114	0.131	0.190	0.148	0.253	0.158

Table 3. Production (kg of fresh material) sampled per block of the Granada population.

SAMPLING	2008	2009	2010	2011	2012	2013	2014	mean
BLOCK I INNER	13.80	12.22	12.00	14.64	11.92	15.42	14.40	13.49
BLOCK II INNER	18.25	28.90	15.50	15.60	15.87	17.55	18.30	18.57
BLOCK III INNER	18.50	38.10	18.50	16.88	13.26	17.98	20.20	20.49
BLOCK I OUTER	12.00	12.54	13.50	16.90	14.40	18.58	14.40	14.62
BLOCK II OUTER	22.50	14.44	17.00	19.34	18.20	22.24	15.20	18.42
BLOCK III OUTER	21.50	51.70	19.50	25.48	22.54	23.58	19.30	26.23
BLOCK I TOTAL	25.80	24.76	25.50	31.54	26.32	34.00	28.80	28.10
BLOCK II TOTAL	40.75	43.34	32.50	34.94	34.07	39.79	33.50	36.98
BLOCK III TOTAL	40.00	89.80	38.00	42.36	35.80	41.56	39.50	46.72
TOTAL SAMPLE	**106.55**	**157.90**	**96.00**	**108.84**	**96.19**	**115.35**	**101.80**	**111.80**

Table 4. Total essential oil (l) from the Granada population.

	HARVEST	2008	2009	2010	2011	2012	2013	2014	mean
BIOMASS	BLOCK I	296	307	260	345	286	388	353	319
	BLOCK II	306	318	322	411	270	410	349	341
	BLOCK III	294	359	358	416	311	402	385	361
	TOTAL CUTTING	**896**	**984**	**940**	**1172**	**867**	**1200**	**1086**	**1021**
	SAMPLING	107	158	96	109	96	115	102	112
	TOTAL HARVEST	**1002**	**1141**	**1036**	**1281**	**963**	**1315**	**1188**	**1132**
	ESSENTIAL OIL	**1,703**	**1,000**	**1,081**	**1,540**	**1,557**	**1,770**	**1,830**	**1,497**

Table 5. Total essential oil (l) from the Granada population.

	Granada population																		
	Block I						**Block II**						**Block III**						**Total**
Rows	1	2	3	4	5	6	1	2	3	4	5	6	1	2	3	4	5	6	
2011	0	0	0	0	0	0	0	1	0	0	0	2	0	0	0	1	0	0	4
2012	1	0	1	1	0	1	0	2	1	0	0	3	0	1	0	2	0	1	14
2013	1	0	1	1	1	2	6	7	2	2	0	3	0	2	0	2	0	1	31

b) Teruel population (Tables **6-9**)

Fresh Material

The sample mean for all production years was 1.77 kg/plant with a sample standard deviation of 0.43 kg/plant which, considered jointly with all the data collected, puts the end points of the confidence interval based on the Student's t-distribution for the mean, at 95% of the confidence level between 1.37 and 2.17 kg/plant. This means that the sample cutting figures for production year two (2009) are outside of the interval (the same goes for the Granada population) with an extraordinary production peak with per-hectare production in excess of 30 tons. The range is greater, 1.20, between the minimum of 1.49 kg to the maximum of 2.70 kg/plant. The sample median is 1.57 kg/plant and 20,888 kg/ha, lower than the mean because it is a more robust measurement tending more towards the middle and is more stable than the mean.

All measurements corresponding to the two populations and, naturally, the mean value of all the years is within the sample range, which means that the sample is representative of the population. The sample mean is also representative of the sample with a variation coefficient of 0,24.

Sample production per hectare without the 2009 outlier figure is 21,515 kg/ha. The mean, which is not very robust due to some extreme data, is 23,581 kg/ha, approximately two tons/ha higher.

The harvest mean is 1.63 kg/plant and 21,696 kg/ha, 8% less than the sample mean. Unlike the sample data, there is no outlier figure for 2009 and it is therefore safe to say that this higher figure is exclusively the result of greater development of some of the plants in rows 3 and 4 and is not representative of the year. The dispersion of the population is substantially reduced, with a standard deviation of 0.17 kg/plant and 2.26 kg/ha giving a variation of approximately 0.1. It can be concluded that during the productive years the population is extremely homogeneous and therefore the mean is representative of the entire population.

Dry Material

Returning to the Teruel population, dry biomass is estimated from harvest and the ratio of fresh and dry biomass is calculated.

The mean figure for the population is 0.83 kg/plant and 11.04 kg/ha, with a median of 0.84 kg/plant and 11,148 kg/ha. The similarity between these two measurements suggests low dispersion. This dispersion measured by the standard deviation comes to 0.09 kg/plant and 1,225 kg/ha and, the range, is 0.32 kg/plant and 4,203 kg/ha. The interquartile range would be a more appropriate measure of the dispersion than range but because of the small population size, the quartiles are found distorts their significance.

The low degree of dispersion translates into a variation coefficient of 0.11 which again suggests that the variables of this study are homogeneous and that measurements tending towards the middle are representative.

Essential Oil

The essential oil figures of the Teruel population were obtained from the total harvest of the wormwood crop. While there are sample data on essential oil distillation in some cases, the scant data available and the heterogeneous nature of sample distillation each year make it impossible to compare samples over the experimental period.

The essential oil variable is obtained from harvest data. On average, each plant produced 2.85 ml resulting in 39.2 liters per hectare. The median provides very similar values: 2.79 ml per plant and 39.2 liters per hectare. The dispersion is calculated at 0.57 ml/plant and 8.9 liters/ha resulting in variation coefficients in the 0.20 to 0.22 range confirming the representativeness of the measurements.

The confidence interval limits for the mean per plant and per hectare are 2.32 and 3.38 ml/plant and 31.01 and 47.47 l/ha.

Finally, the shape of the distribution is determinate by an asymmetry measure of 0.6 and 0.7, and a kurtosis measure of -0.9 and -0.2 per plant and hectare respectively. In both cases the populations demonstrate positive asymmetry with greater data dispersion for higher levels of production and very concentrated data at lower production. In other words, production will most likely be in the vicinity of these figures. Distribution is platykurtic and more rounded than normal which means that extreme values are more likely and central values less likely than in a normal distribution.

Plant Mortality

Presenting similar results to Granada population, Teruel population also suffered high mortality in 2012 and 2013. Table **9** shows mortality in Teruel population.

Mortality was lower than for the Granada population but was still significant and needs to be taken into account. Mortality was zero up to 2011. However, two plants were lost in 2011 that did not affect the sample population and this would not have been significant had it not been for what happened later: six plants were lost in 2012 and 12 more were lost in 2013, a total of 20 plants.

Table 6. Teruel Population. General data.

	2008	2009	2010	2011	2012	2013	2014	mean
Production per plant (kg)	1.61	1.76	1.60	1.91	1.40	1.63	1.47	1.63
Production per hectare (kg/ha)	21,494	23,530	21,355	25,446	18,707	21,674	19,665	21,696
Essential oil per plant (l)	2.81×10^{-3}	2.10×10^{-3}	2.53×10^{-3}	3.57×10^{-3}	3.79×10^{-3}	2.36×10^{-3}	2.80×10^{-3}	2.85×10^{-3}
Essential oil per ha (l/ha)	37.51	28.00	33.79	47.66	55.45	31.53	40.76	39.24
Essential oil per production (l)	0.174	0.136	0.172	0.203	0.296	0.157	0.249	0.198

Table 7. Production (fresh biomass) sampled per block of the Teruel population.

SAMPLING	2008	2009	2010	2011	2012	2013	2014	mean
BLOCK I INNER	14.00	11.96	13.50	14.30	13.46	12.58	11.10	12.99
BLOCK II INNER	16.00	30.80	15.50	15.00	11.48	14.50	14.00	16.75
BLOCK III INNER	14.20	29.70	13.00	15.04	12.32	9.58	20.80	16.38
BLOCK I OUTER	16.00	13.22	13.50	16.12	19.26	16.76	11.90	15.25
BLOCK II OUTER	20.50	37.83	18.50	24.58	17.20	16.88	15.40	21.56
BLOCK III OUTER	21.50	38.40	20.00	26.42	17.76	19.38	18.90	23.19
BLOCK I TOTAL	30.00	25.18	27.00	30.42	32.72	29.34	23.00	28.24
BLOCK II TOTAL	36.50	68.63	34.00	39.58	28.68	31.38	29.40	38.31
BLOCK III TOTAL	35.70	68.10	33.00	41.46	30.08	28.96	39.70	39.57
TOTAL SAMPLE	**102.20**	**161.91**	**94.00**	**111.46**	**91.48**	**89.68**	**92.10**	**106.12**

Table 8. Production of biomass (kgs) and essential oil (liters) of the Teruel population.

	2008	2009	2010	2011	2012	2013	2014	mean
Biomass (Block I)	403	365	348	418	314	364	377	370
Biomass (Block II)	330	399	368	457	343	383	339	374
Biomass (Block I)	356	377	372	422	287	363	280	351
Sample	102	162	94	111	91	90	92	106
TOTAL BIOMASS	**1,088**	**1,141**	**1,088**	**1,297**	**944**	**1,110**	**996**	**1,095**
TOTAL ESSENTIAL OIL	**2.08**	**1.55**	**1.87**	**2.15**	**2.46**	**1.75**	**2.07**	**1.99**

Table 9. Plant mortality (number of plants), Teruel population.

	Teruel Population																		
	Block I						**Block II**						**Block III**						**Total**
Rows	1	2	3	4	5	6	1	2	3	4	5	6	1	2	3	4	5	6	
2011	0	0	0	0	0	0	0	1	0	0	0	0	1	0	0	0	0	0	2
2012	1	0	0	0	0	0	0	2	0	0	0	0	2	1	1	0	1	0	8
2013	3	1	1	1	0	0	0	3	0	1	0	2	2	2	3	0	1	0	20

Variable Analysis and Year of Study

For the year 2008 production results of Teruel population were unusually high considering it was the first production year. Both per-plant and per-hectare biomass production was the fourth largest of the six years of the experiment. The figures are even higher for the total amount of essential oil in that first year. For the Granada population, biomass production results were as expected with low production (Tables **2** and **6**).

In 2009, fresh biomass production increased for both populations in comparison with the first year, with the highest per-plant and per-hectare production except for 2011, and above the respective mean values (Tables **2** and **6**). Despite these results, 2009 was the year with the lowest essential oil production for both populations (Tables **4** and **7**).

2011 was the second most productive year for the Teruel population and the best year for the Granada population (Tables **2**, **4**, **5** and **7**). Production per sample plant is very similar in the two varieties, both above 1.80 kg per plant and 24,000 kg per hectare (Tables **3** and **6**).

In 2012 production of both varieties was the lowest of the six years (Tables **2**, **4**, **5**, **7**). In the case of Teruel it was indisputably the lowest, and in the case of Granada sample measurements per plant and per hectare were slightly higher than those of 2010 (Tables **3** and **6**), but overall production was also clearly the lowest.

Data variability rose in 2013. Sample data for Teruel were the lowest of the experiment (Tables **3** and **6**). However, in terms of total production, that year ranked third for this population. In contrast, for the Granada population it was the most productive year of the experiment both in terms of sample and overall data (Tables **2** and **5**). Hence, differences between the two populations were significant that year, both in terms of the sample and, to a lesser degree, the overall plant population. The sample production figures for Granada were nearly 30% higher than those of Teruel. This difference diminishes to less than 20% in the case the total harvest.

In 2014, dry and fresh production of population raised again for the Granada population over average. Essential oil production was over average too. For Teruel population, dry and fresh production were low, but even for Teruel, essential oil production was over average and the third year over 40 liters per hectare. On the beginning the timeline was thought about 5 or 6 of production years, but according to the data of this last year the production horizon could be more extent.

Biomass Variable

<u>2008 Results</u>

Sampled biomass production exhibits similar distribution for the two populations when broken down into blocks, with a more pronounced mode in block II in the

Teruel population and lower production in Block I. Production dispersion per block is higher in the Granada population. Nevertheless the average production of approximately 35 kg per block and 1.75 kg per plant are similar for the two populations with a slight increase in the Granada population (Figs. **2** and **3**).

If we look at total production for the Teruel population, we find that the sample distribution is completely heterogeneous (Diag. **4**). Fresh biomass production is highest in block I and lowest in block II. Production per block was between 330 and 405 kg, giving a total production that year of 1087.5 kg which, together with sampled production, gives a total of 1189.7 kg.

Total production in the Granada population was lower than that of Teruel with results that did not coincide with sample measurements; nearly 20% lower production of fresh biomass for the Granada population (Fig. **5**).

2009 Results

Biomass production by blocks was higher than the previous year but was not the same for all blocks. This production increase puts the average over 50 kg per block with a per-plant average practically a kilo higher than in 2008, *i.e.* a 50% increase (Figs. **2** and **3**).

Total fresh biomass production that year in Teruel was 1140.5 kg which, together with the sampled plants gives a final production of 1302.4 kg for. In the Granada population, production in blocks was higher than the previous year. Total production was 983.5 kg which, together with the sampled output, gives a total of 1,141.4 kg for that year (Figs. **4** and **5**).

2010 Results

Similar values were obtained for biomass, although below those of the first year thus confirming the anomaly of 2009. Data dispersion is lower in the Teruel population than in Granada. The averages are similar for the two populations in the vicinity of 31.5 kg and plant production at approximately 1.6 kg. Once again, measurements are slightly higher for the Granada population (Diags. **2** and **3**).

Harvest that year for the Teruel population totaled 1088 kg which, together with the fresh biomass sampled, comes to a total of 1182 kg. In the Granada population, 940 kg of fresh biomass was obtained which, together with the sampling material, gives a total of 1036 kg (Diags. **4** and **5**).

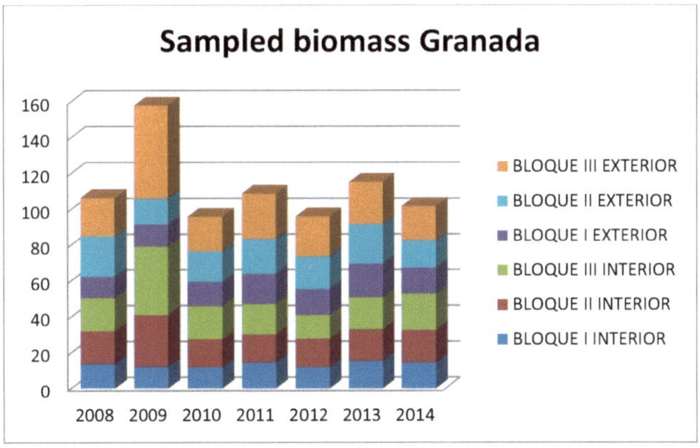

Diag. (2). Sampled fresh biomass, Granada, by blocks (kg).

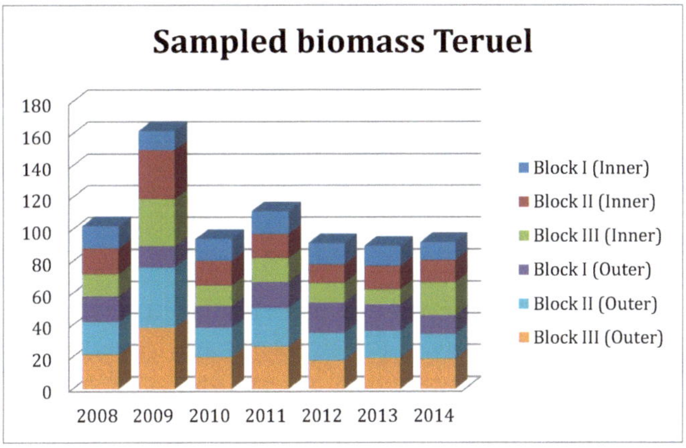

Diag. (3). Sampled fresh biomass, Teruel, by blocks (kg).

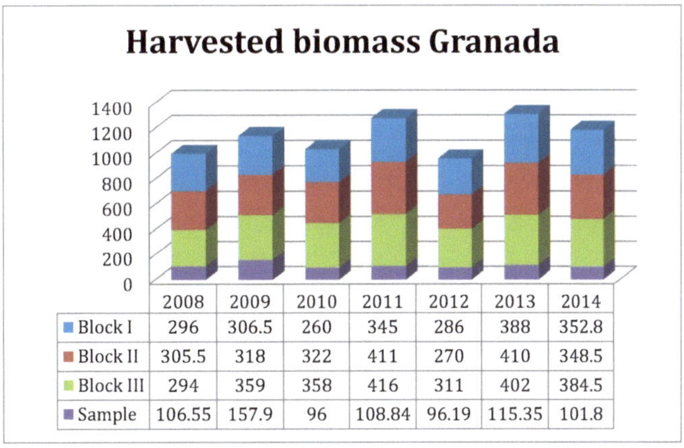

	2008	2009	2010	2011	2012	2013	2014
Block I	296	306.5	260	345	286	388	352.8
Block II	305.5	318	322	411	270	410	348.5
Block III	294	359	358	416	311	402	384.5
Sample	106.55	157.9	96	108.84	96.19	115.35	101.8

Diag. (4). Harvested fresh biomass, Granada, by blocks (kg).

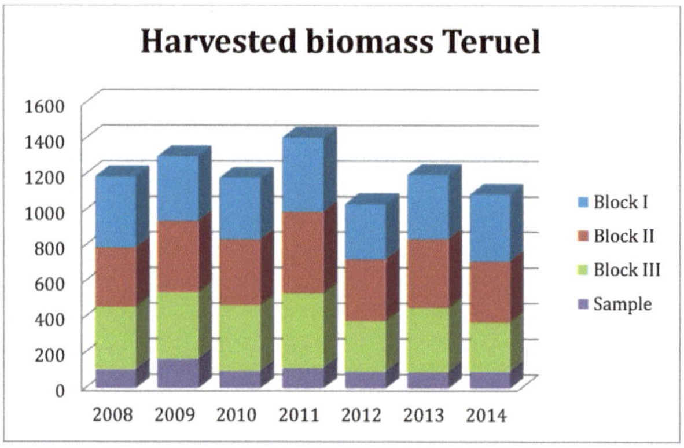

Diag. (5). Harvested fresh biomass, Teruel, by blocks (kg).

2011 Results

Per-block and per-plant averages are again very similar, approximately 37 kg per block and 1.8 kg per plant. This time the sample measurements are slightly higher for the Teruel population. Production for the Teruel population was 1297 kg which, together with the sampled material, brings total production to 1408.5 kg for that year (Diags. **2** and **3**).

In the Granada population, distribution by block continued to rise. This year, both sample and harvest data are lower for the Granada population. Production by block totals 1172 kg which, together with the sampled amount, gives a total production of 1280.8 kg (Diags. **4** and **5**).

2012 Results

The average per block is approximately 31 kg and per plant 1.55 kg. Both figures are slightly higher than the production of the Granada population (Diags. **2** and **3**).

Harvest production was below 1000 kg (944), a surprising figure for the Teruel population. The total (harvest + sample) comes to 1035.5 kg. Harvest production for the Granada population was 867 kg and harvest + sample comes to 963.2 kg (Diags. **4** and **5**).

2013 Results

This year the Granada population exceeded the mean by producing 38 kg per block. The same is true for per-plant production; while the Teruel population was under 1.5 kg, the Granada population was just below 2 kg per plant (Diags. **2** and **3**).

Total production was 1200 kg of biomass making it the second worst year. Production reached maximum levels in the Granada population, with the exception of block III. This increase in production was similar for all blocks. The total was 1315.4 kg, significantly higher than other years and, for the first time, higher than the Teruel population (Diags. **4** and **5**).

2014 Results

In the last year of production, as in the previous one, a significant number of plants were lost, more in the Granada than the Teruel population. Although the experiment was originally designed to conclude in 2013, another year was added because good production was expected. The production of fresh and dry biomass was slightly below the overall average for the study but, surprisingly, both populations produced volumes of essential oil that were close to or above the average. Per-hectare production for the Granada population was over 21 tons of fresh and 10 tons of dry material. Similar results were achieved by the Teruel population with 20.4 tons/ha of fresh biomass, nearly 60% of which was converted into dry material thus reaching 11.65 tons/ha of dry material (Diags. **4** and **5**).

Essential Oil Variable

Diags. (**7** and **8**) show the evolution of essential oil production in Teruel and Granada.

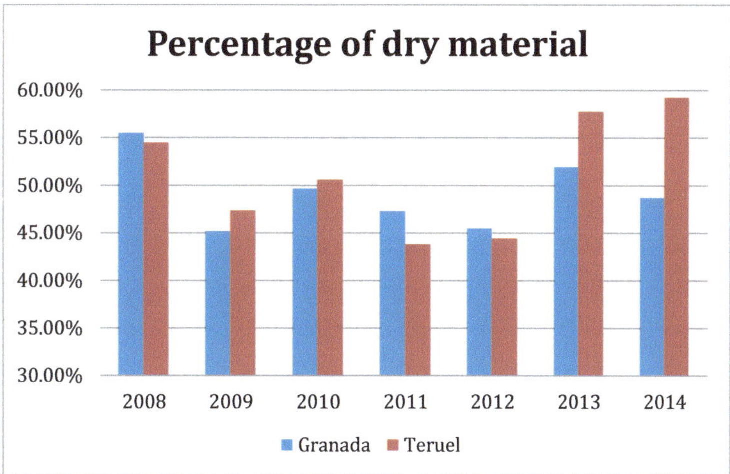

Diag. (6). Comparison of dry material, Granada and Teruel (%).

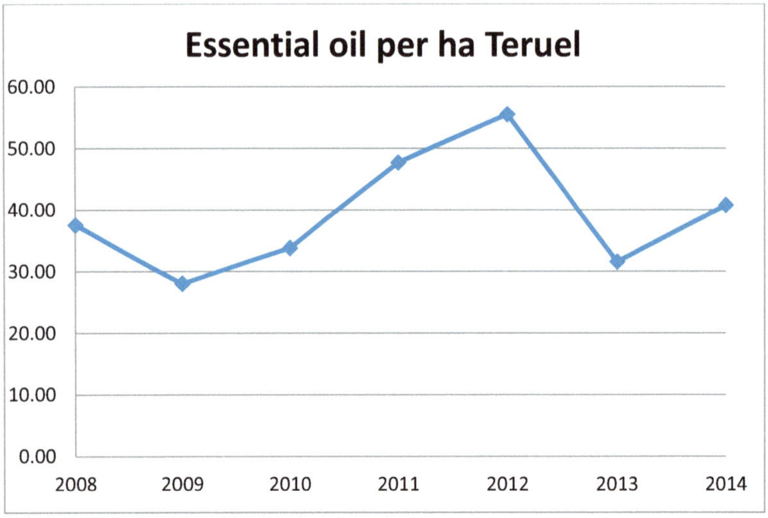

Diag. (7). Evolution of essential oil production, Teruel (1).

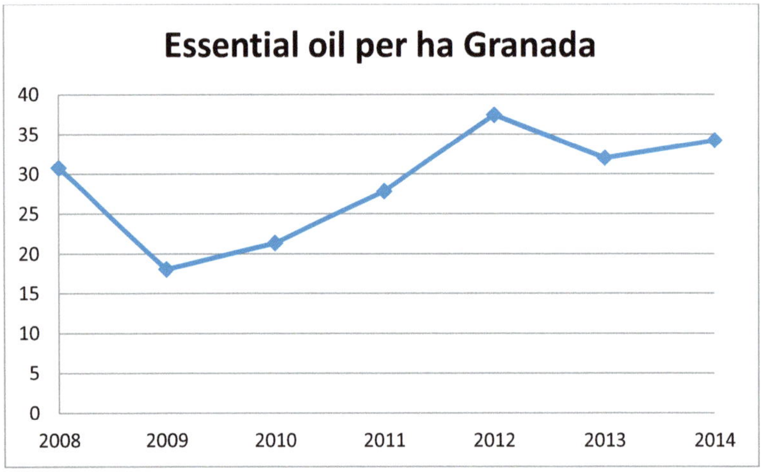

Diag. (8). Evolution of essential oil production, Granada (1).

2008 Results

Essential oil production from the Teruel population was very high during the first year and followed 2011 and 2012 as the third highest (Table **5**) but ranked second in terms of per-hectare production (Diag. **6**) and per 100 kg of harvested material and was the best year for production per plant.

2009 Results

As already mentioned regarding the production of fresh biomass this year, despite the large amount of biomass produced, essential oil as a percentage of the total production and per-plant, per-hectare and per-100 kg of biomass, was the lowest of the entire experiment for both populations (Table **6**). The difference in the production of essential oil is significantly lower. This is clearly illustrated in the essential oil production diagrams.

2010 Results

The decline in biomass production for both populations also translates into lower amounts of essential oil. The oil figures for this year are below the average levels obtained throughout the experiment.

2011 Results

Both this year and the previous year there was a clear correlation between the amount of fresh biomass harvested and essential oil produced. It was the second best year for essential oil production in the case of the Teruel population and the best year for Granada. Evens so, production in the two populations is significantly different, Teruel producing over 30% more than Granada.

Essential oil production from the Teruel population was nearly 40 l/ha while that from Granada was slightly below 30 l/ha. In terms of oil per 100 kg of biomass, Teruel out-produced Granada by over 50%.

2012 Results

In contrast to the amount of fresh biomass harvested this year, 2012 was the best year for essential oil production (followed by 2011). Again, there was a significant difference in production between the two populations with Teruel out-producing Granada. Teruel produced over 50 liters per hectare and just under 0.3 liters per 100 kg of fresh biomass. Neither of these figures was matched any other year.

2013 Results

Essential oil production in the Teruel population only exceeds that of 2008, which is no surprise as that year ranked sixth. In contrast, it was the second best year for essential oil production in the Granada population.

2014 Results

34.2 liters of essential oil per hectare was obtained by Granada population, the second highest amount of the study. Teruel population produced 40.8 l/ha of essential oil, above the study average on both populations. Both varieties reach maximum essential oil production on the fifth year of production (2011), but this last year is an increase respect previous and one of the more productive years.

Anova Procedures

A statistical test to compare the equality of variances was calculated (Tables **10** and **11**). The value of the statistical F with 6 and 6 degree of freedom is 1.20945 with a p_value of 0.8233 (bilateral), then it can be considered the variances are the same. After that is comparing the equality of the means of the two ecotypes considering the equality of variances.

Table 10. Biomass general data of ecotypes (kg/ha)

Variable	N	Average	Desv. stand.
Biomass Teruel	7	21,696	2,259
Biomass Granada	7	20,457	2,485

Table 11. Oil production- general data of ecotypes (l/ha)

Variable	n	Average	Desv. stand.
Oil production -Teruel	7	39.2412	9.6126
Oil production- Granada	7	28.7976	6.9415

The t-Test with 12 degrees of freedom is 0.975 and a p_value of 0,349. It shows no significant differences in yearly biomass per hectare between both ecotypes.

Repeating ANOVA test for oil production variable the value of the statistical F to compare the equality of variances was of 1.9177 with a associated *p*_value of 0,448; then in this case it can be considered the variances are the same as the previous case. Anova test is comparing the equality of means of the two ecotypes considering the equality of variances.

The t-test with 12 degrees of freedom is 2.33 and a *p*_value of 0.038 (bilateral). It shows no significant differences in yearly oil production in liters per hectare between both ecotypes. Production of essential oil is significance superior in the case of Teruel population.

It has to be careful with the results of the Anova test because there were differences between years with the distillation procedure and the high mortality rate. It is important to bear in mind the small number of observations, *i.e.* the number of crop seasons, as a limit to the robustness of the results or conclusions. To ensure the results of this experiment, we would need to study the variables with enough number of individual plants or subgroups in a larger plot with a minimum of thirty observations to have a sufficient degree of freedom and to employ the central limit theorem.

All the central measures of the variables studied are sufficiently representative, in terms of point and interval estimate. Therefore, by using the mean, or where there is an extreme data distribution of the median, total distribution can be simplified.

Sample measurements were significantly representative of the total harvest. Only 5% sample was able to provide an accurate estimation to the total harvest. Hence, accurate estimates can be made in cases where we need to know total production without having to harvest or distill the whole crop. Dry material is the most homogeneous variable of all those studied and is therefore the best to estimate production based on past years. It is also the easiest variable to predict with the lowest average error.

There is not a very clear correlation between the yield of fresh biomass and the amount of essential oil obtained from it. The 2009 Granada population merits special mention as biomass production, being unusually high that year, with an unusually low essential oil production. This could indicates that beyond a certain optimal size, additional development of the wormwood plant is at the expense of essential oil production. Further study is needed to support this theory.

At Teruel population, there is a quadratic relationship between the production of dry material and essential oil.

On average, harvest figures for all variables were higher for Teruel population. Therefore, if the aim is to obtain biomass, dry material or essential oil, Teruel population is the one to be chosen. It is also important to point out that the mortality rate for Teruel population was approximately 30% lower than in Granada population, another good reason to choose Teruel one. For example, in order to obtain 100 liters of essential oil from wormwood, 3.47 hectares of Granada population would have to be planted compared to 2.55 hectares of Teruel population.

The ratio between fresh biomass produced and essential oil can be extrapolated from the main results (Diag. **9**). As there is no linear relationship between the two variables, we can only just estimate that for standard production of between 20 and 25 thousand kilos per hectare, the essential oil obtained would almost surely be greater than 30 liters per hectare. This is vital information needed to establish a standard production to obtain a certain minimum amount of essential oil.

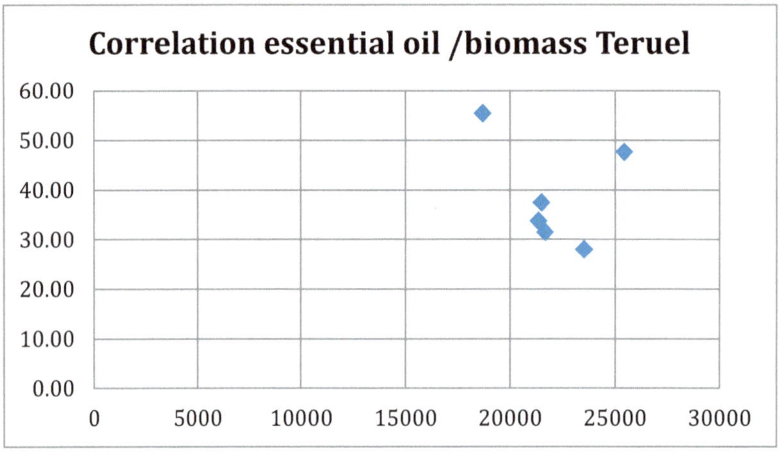

Diag. (9). Correlation between essential oil production and fresh biomass for the Teruel population.

Lastly, based on the estimates set out in Table **12** and the total production of essential oil per hectare in the Teruel population (274.69 liters at an average of 39.24 l/ha/year), 2.55 hectares would need to be planted to obtain 100 liters/year. Based on the minimum production observed in the study, in order to ensure those 100 liters/year, 3.6 hectares would have to be planted.

Table 12. Costs in euro/hectare for 7 production years.

Preparatory work	80
Soil analysis	70
Crop planting	250
Preparatory work and crop planting	**400**
Weeding	705
Fertilizing	994
Harvest	2,590
Maintenance work and harvesting	**4,289**
Indirect costs (taxes)	848
FULL PROD COST	**5,937**

Note: If necessary, a drip irrigation system would have to be added to the above table. This would entail an additional cost of EUR 160 per year (25-year uniform amortization) or EUR 80 per year (50 year uniform amortization). The cost of oil extraction must likewise be added and may range from EUR 3.00-4.50/liter of essential oil.

The following graph shows the total amount of essential oil obtained from distillation:

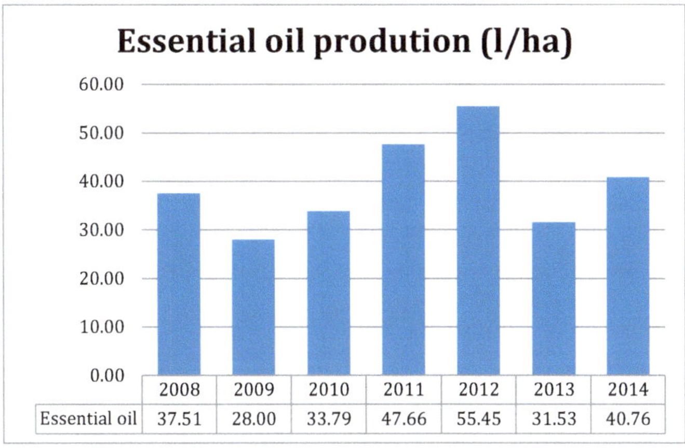

Diag. (10). Total essential oil obtained from all crop years (l/ha).

Total per-hectare essential oil production from all years of the experiment was 274.69 liters, an average of 39.24 liters per hectare and year. Therefore, 2.55 hectares are needed to produce 100 liters/year. Based on the minimum production observed in the study, in order to ensure those 100 liters/year, 3.6 hectares would have to be planted (Diag. **10**).

Based on the best stable production results of the Teruel population and its lack of thujone in analyses detailed in the next chapter, a plant variety patent registration process was initiated at the Community Plant Variety Office in 2011. The result is the Candial variety ® by EU decision No. 36714 of 27 January 2014.

CONFLICT OF INTEREST

The authors declare no conflict of interest, financial or otherwise.

ACKNOWLEDGEMENTS

Declared none.

REFERENCES

[1] Burillo J. Investigación y experimentación de plantas aromáticas y medicinales en Aragón: cultivo, transformación y analítica. 2003.

[2] Font Quer P. Plantas Medicinales. Rl Dioscórides renovado. Barcelona: Editorial labor 1981.

[3] Höld KM, Sirisoma NS, Ikeda T, Narahashi T, Casida JE. Alpha-thujone (the active component of absinthe): gamma-aminobutyric acid type A receptor modulation and metabolic detoxification. Proc Natl Acad Sci USA 2000; 97(8): 3826-31.

[http://dx.doi.org/10.1073/pnas.070042397] [PMID: 10725394]

[4] Villar L, *et al.* Plantas Medicinales del Pirineo Aragonés y demás tierras oscenses. 2nd ed., CSIC-DPH 1987.

[5] Burillo J, García-Vallejo MC. Investigación y Experimentación de Plantas Aromáticas y Medicinales en Aragón Cultivo, transformación y analítica. 2003.

[6] Perejo I, *et al.* Comparison between the radical scavenging activity of six distilled and non-distilled Mediterranean herbs and aromatic plants. J Agric Food Chem 2002; 50: 81-6. [PMID: 11754547]

[7] Gepts P. Crop domestication as a long-term selection experiment. In: Janick J, Ed. Plant Breeding Reviews, 24, part 2. John Wiley and Sons, INC 2004.

[8] Pickersgill B. Domestication of plants in the Americas: insights from Mendelian and molecular genetics. Ann Bot 2007; 100(5): 925-40. [http://dx.doi.org/10.1093/aob/mcm193] [PMID: 17766847]

[9] Davies P. Estudios en domesticación y cultivo de especies medicinales y aromáticas nativas. Uruguay. Serie FPTA-INIA 2004; 11: 15.

[10] Umpiérrez ML, *et al.* Essential oils from Asteraceae as potential biocontrol tools for tomato pests and diseases. Phytochem Rev 2012; 11: 339-50. [http://dx.doi.org/10.1007/s11101-012-9253-5]

[11] Carnat AP, *et al.* Cis-chrysanthenol, a main component in essential oil of Artemisia absinthium L. growing in Auvergne (Massif Central), France. J Essent Oil Res 1992; 4: 487-90. [http://dx.doi.org/10.1080/10412905.1992.9698115]

[12] Chialva F, *et al.* Chemotaxonomy of wormwood (Artemisia absinthium L.) I. Composition of the essential oil of several chemotypes. Z Lebensm Unters Forsch 1983; 176: 363-6.

[13] Geszprych A, *et al.* Diversity of Wormwood (Artemisia absinthium L.) Growing Wild in Poland in Respect of the Content and Composition of Essential Oil and Phenolic Compounds. Acta Hortic 2011; 123-9. [http://dx.doi.org/10.17660/ActaHortic.2011.925.16]

[14] Günther E. The Essential Oils. History and origin in Plants Production Analysis. New York, USA: Krieger Publishing 1948; 1.

[15] Cordero JM, Córdoba A, Fernández S. Estadística Descriptiva. Madrid: ESIC 2002.

Frontiers in Horticulture, 2017, *Vol. 1*, 163-196

CHAPTER 11

A Case Study of MAPs Production, Uses and Commercialization *Artemisia absinthium* Var. Candial: Extract Characterization and Valorization

Azucena González-Coloma[1,*], Carmen Elisa Díaz[2], Luis F. Julio[1], Jesús Burilo[3] and M. Fe Andres[1]

[1] *Instituto de Ciencias Agrarias, CSIC. Serrano 115-dpdo, 28006Madrid, Spain*

[2] *Instituto de Productos Naturales y Agrobiología, CSIC. Avda. Astrofísico F. Sánchez 2, 38206 La Laguna, Tenerife, Spain*

[3] *Departamento de Ciencia, Tecnología y Universidad, Centro de Investigación y Tecnología Agroalimentaria de Aragón, Gobierno de Aragón, Avda. Montañana, 930, Zaragoza, Spain*

Abstract: Two thujone free Spanish populations of *Artemisia absinthium* from Teruel and Sierra Nevada (Spain) were submitted to a domestication process. The domestication resulted in lower chemical variation and higher biomass and essential oil yields, allowing for the registry of a new domesticated variety (® Candial). The main components of the new variety essential oil extracts were *cis*-epoxyocimene, chrysanthenol, chrysanthenyl acetate, *trans*-caryophyllene and linalool. The presence of 1.8-cineol and camphor correlated with the domestication level of *A. absinthium*. CO_2 supercritical fluid extraction, improved the yield of (*Z*)-epoxyocimene, chrysanthenol, (*5Z*)-2,6-dimethylocta-5,7-diene-2,3-diol and *trans*-chrysanthenyl acetate. The development of extraction processes with supercritical CO_2 allowed for the enrichment of desired bioactive compounds in *A. absinthium* extracts. The superctitical extracts showed strong antifeedant and insecticidal effects against *Spodoptera littoralis*, indicating that for insect control purposes supercritical CO_2 extraction should be used. The semi-industrial vapor-pressure extracts (VPs) showed strong antifungal effects. (-)-*cis*-chrysanthenol has been identified as the chemical indicator of the antifungal quality of VP extracts from *A. absinthium* (® Candial). These VP extracts also showed a high nematocidal activity against animal parasitic nematode *Trichinella spiralis* without being cytotoxic to macrophage cells. These oils showed activity on the protozoan parasites *Trypanosoma cruzi* and *Trichomonas vaginalis*. The antiprotozoal activity of the VP oil was related with the presence of *trans*-caryophyllene and dihydrochamazulene. EOs from cultivated *A. absinthium* could be considered as potential candidates for development of new antifungal and insect antifeedant agents, as well as drugs against helminth and protozoan parasites.

* **Corresponding author Azucena González-Coloma:** Instituto de Ciencias Agrarias, CSIC. Serrano 115-dpdo, 28006 Madrid, Spain; Tel: +34 917452500; E-mail: azu@ica.csic.es

M. Paz Arraiza, A. González-Coloma, J. Burillo and C. Calderón-Guerrero (Eds.)

Keywords: Antifungal, *Artemisia absinthium*, Chrysanthenyl acetate, (-)-*Cis*-Chrysanthenol, Cis-epoxyocimene, Dihydrochamazulene, Domestication process, Insect antifeedant, Linalool, Supercritical CO2, Trans-caryophyllene, *Trichinella spiralis*, *Trichomonas vaginalis*, *Trypanosoma cruzi*.

INTRODUCTION

The genus Artemisia belongs to the family Compositae (Asteraceae) and consists of about 500 species distributed through the world. *Artemisia absinthium* L. is an aromatic and medicinal plant of ethnopharmacological interest [1, 2]. The composition and biological effects of the essential oil (EO) of *A. absinthium* have been widely studied. *A. absinthium* EOs had antimicrobial [3 - 5] and antiprotozoal effects against *Leishmania aethiopica* and *L. donovani* [6]. In addition, thujone-rich oils have been shown to have acaricidal [7], insecticidal [8 - 10] and fungicidal effects [10] and myrtenol-rich oils repelled fleas, flies, mosquitoes [11] and ticks [12].

Among the major EO components reported are α and β-thujone [10, 13 - 15], myrcene, *trans*-sabinyl acetate [15 - 19], β-pinene [9], 1.8-cineol [9, 20], camphor [6], *cis*-epoxyocimene [14], chrysanthenyl acetate [9, 14, 15], sabinene [9, 15], myrtenol [11, 12], bornyl acetate [22], artemisia ketone, linalool, hydrocarbon monoterpenes [9], sesquiterpene lactones [23, 24] isoascaridol [25] and mixtures of some of these components [13 - 15, 25] depending on the plant origin.

A. absinthium is abundant in the mountains of Spain as a ruderal species and it is used as a medicinal remedy. There are seven chemotypes described in the Iberian Peninsula and some of them are thujone-free [26]. Therefore, the use of *A. absinthium* based on the collection of wild populations can result in variable compositions of the extracts.

As part of an ongoing project on the valorization of native plants for the sustainable production of botanical biopesticides, a Spanish population of wormwood from Teruel was introduced in a domestication program [24, 27 - 30]. The essential oils were characterized by the presence of *cis*-epoxyocimene, chrysanthenol, and chrysanthenyl acetate. These oils repelled reduviid insects and showed moderate effects against fungal pathogens [30, 31]. Furthermore, supercritical fluid extracts of this *A. absinthium* population showed an improvement in the yield of several mono- and sesquiterpenes and were more antifeedant than the traditional extracts (hydrodistilled and organic) against phytophagous insects [24, 29]. Selected germplasm from the pre-domesticated *A. absinthium* population (Teruel) entered a second field cultivation cycle for agronomic stabilization and further valorization. Additionally, wild germplasm from a Sierra Nevada (Spain) population of *A. absinthium* entered the same cultivation process for pre-domestication and comparison purposes [32]. The domestication process resulted in the registration of a plant variety (®Candial) and the development of a new wormwood-based biopesticide.

The agronomic aspects of the domestication process have been described in detail in Chapter 10. In this chapter we will describe the chemical and biological parameters of importance in the domestication and valorization of Spanish wormwood.

Cultivated Spanish Wormwood Essential Oil

Pre-Domesticated Wormwood Essential Oil

Plant material for field cultivation was selected from a thujone-free wild population (T0) growing in Teruel (Spain). The individuals for field cultivation were obtained from seeds and cultivated in two locations (Ejea de los Caballeros, E; and Teruel, T; Spain) to give the pre-domesticated population T1 [28]. Flowering samples from 30 randomly selected plants were collected yearly and processed for essential oil (EO) hydrodistillation by a Clevenger apparatus. The EOs were analyzed for five consecutive years (2002-2006: population T1, crops e1-e5 and t2-t5, Table **1**) [27].

Table 1. ***Arthemisia absinthium*** **samples of different origins (Sierra Nevada, SN, Teruel, T; Villacampa, V), produced under different conditions (wild, field and controlled environment) in different locations (Ejea and Teruel) for five years.**

Origin	Year	Cultivated					Wild
		Growth chamber	Greenhouse	Aeroponic	Field		
					Ejea[a]	Teruel[b]	
SN[c]	1	SNC	SNI	SNA			
T[b]	1				T1e1	na	
	2				T1e2	T1t2	
	3				T1e3	T1t3	
	4				T1e4	T1t4	
	5				T1e5	T1t5	
V[d]	-						V

[a]Ejea de los Caballeros (Zaragoza): 42°07’16”N, 1°08’20”O, 398m
[b]San Blas (Teruel): 40°21’30”N, 1°10’45”O, 923m
[c]Sierra Nevada (Granada): 37°05’33”N, 3°23’04”O, 2083m
[d]Villacampa (Huesca): 42°30’32”N, 0°21’21”O, 759m.

Another wild population (Sierra Nevada, SN) was obtained from Sierra Nevada National Park (Granada) and cultivated in a growth chamber (C), aeroponic chamber (A) and green house (I) for one year [29]. Aerial parts were collected for hydrodistillation (SNC, SNI and SNA samples respectively). For comparison purposes, a different wild population V (Villacampa, Huesca, Aragón, 2006) was also included in the study. A summary of all the samples studied is shown in Table **1**. Tables **2** and **3** show their chemical composition.

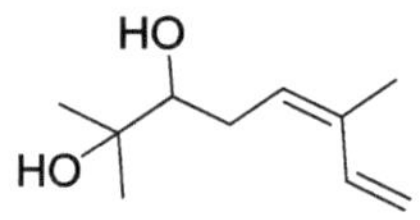

Fig. (1). (Z)-2,6-dimethylocta-5,7-dien-2,3-diol (4).

Table 2. Chemical composition of the different *A. absinthium* samples (abundance: % area).

Compound	T1, Ejea					T1, Teruel				SN0			V
	e1	e2	e3	e4	e5	t2	t3	t4	t5	SNC	SNI	SNA	Wild
α-pinene	0.4	2.01	0.5										
1.8-cineol	1.02		2.5					1.2	0.8				0.6
linalool (**5**)	2.6	4.7	6.1		1.6	2.0	4	2.4	4.4			0.4	
cis-epoxyocimene (**1**)	22.8	59.9	41.8	1.1	2.0	30.5	31.0	44.8	4.3	7.8		75.1	45.7
Trans-epoxyocimene	2.51		4.0			2.3	2.1	2.7					3.3
camphor (**6**)		6.9							1.9				
chrysanthenol (**2**)	10.1	13.2	12.9	9.3	19.3	5.9	8.1	4.4	17.0	12.3	21.4	10.6	11.1
linalyl propionate			1.04						6.2				
linalyl acetate			4.2										
bornyl acetate										15.1			
chrysanthenyl acetate (**3**)	9.8	2.2	11.6	2.4	4.4	10.5	17.2	16.8	19.4		1.5	2.5	32.2
(*Z*)-2,6-dimethylocta-5,7-die--2,3-diol (**4**)				30.0	54.6								
thymol						10.0							
cis-linalool oxide (furanoid)	34.3	6.3	0.5			2.9	24.1	12.2	16.7		54.1		
trans-linalool oxide (furanoid)	1.3						1.2						
caryophyllene (**9**)	1.2	2.0	1.3			3.4	1.3	2.0	2.9	5.7	1.3	1.8	
β-selinene	0.5	1.7	1.9			1.0	0.7	0.5	1.2			0.7	
caryophyllene oxide (**11**)	0.4	1.2	1.1	2.6	5.4		0.5	1.3	3.7		4.3	0.7	

Table 3. Chemical composition of essential oils (Clevenger Hydrodistilled, HDs) of the different *A. absinthium* samples (abundance: % area). Values are expressed as average (range) of individual samples from three years (2008-2010). T, Teruel population; SN, Sierra Nevada population; T2, domesticated T population; T1, parent T population; SN1, pre-domesticated SN population; SN0, parent SN population.

Compounds	Teruel (2008-2010)		Sierra Nevada (2008-2010)	
	T1	T2	SN0	SN1
1.8-cineol	0.6 (0.0 - 1.2)	2.1 (1.3 - 3.6)	0 (0.0 - 0.0)	0.3 (0.0 - 0.8)
linalool (**5**)	2.8 (1.3 - 4.2)	5.8 (5.5 - 6.1)	3.5 (2.3 - 4.4)	3.1 (2.7 - 3.3)
(-)-(*Z*)-epoxyocimene (**1**)	25.7 (0.0 - 45.9)	23.4 (0.0 - 36.4)	39 (19.7 - 50.9)	35.7 (25.3 - 44.2)
camphor (**6**)	n.d.*	4.2 (3.2 - 4.7)	n.d.	n.d.
(-)-*cis*-chrysanthenol (**2**)	16 (10.4 - 23.0)	19.5 (18.4 - 21.6)	26.6 (19.8 - 37.7)	22.9 (13.2 - 28.9)
chrysanthenyl acetate (**3**)	12.3 (0.5 - 20.4)	2.9 (0.0 - 4.9)	0.4 (0.0 - 1.1)	6.8 (0.0 - 20.3)
(*Z*)-2,6-dimethylocta-5,7-diene-2,3-diol (**4**)	25.9 (0.0 - 53.3)	22 (9.5 - 42.5)	3.6 (0.0 - 7.2)	11.2 (0.0 - 20.8)
trans-caryophyllene	1.6 (1.6 - 1.7)	1.9 (1.3 - 2.3)	2.1 (0.9 - 3.2)	2 (1.7 - 2.4)
sesquiterpene alcohol	2.2 (1.7 - 2.5)	1.7 (0.9 - 2.4)	1.9 (1.7 - 2.2)	1.9 (1.8 - 2.1)

*not detected

The EOs were characterized by the presence of *cis*-epoxyocimene (1), chrysanthenol (2), and chrysanthenyl acetate (3) among others. These compounds have been described in other EOs of this plant [9, 13 - 16]. *cis*-epoxyocimene (1) was the major component of the pre-domesticated T1 crops (Table **2**). Among the SN samples, 1 was the major component of SNA (75.1%), was present in SNC (7.8%), and absent in SNI. The major component of the wild sample V was also 1 (45.7%). Chrysanthenol (2) was present in all of the EOs studied in variable amounts (4.4-21.4%). Chrysanthenyl acetate (3) was present in all the field cultivated samples (T1) (1.5-19.4%). SNI and SNA had low amounts of 3 (1.5 and 2.5% respectively) in contrast to the wild sample V (32.2%). Linalool oxide was found in T1 crops e1-3 (34.3-0.5%) and t2-5 (2.9-24.1%) and was the major component of SNI (54.1%). Among the sesquiterpenes, caryophyllene (9) was found in T1 crops e1-3 and t2-5 (1.2-3.4%). Caryophyllene oxide (11) increased with time for the T1-e samples (0.4 to 5.4%). For the SN0 samples, caryophyllene (9) (1.3-5.7%) was present in all of them [27].

Some compounds were only found in one or two samples (Table **2**). Camphor (6) was present in T1 crops e2 and t5 (6.7 and 1.9%), linalyl propionate in T1 crops e3 and t5 (1.04 and 6.2%), linalyl acetate in T1-e3 (4.2%), bornyl acetate in SNC

(15.1%) and thymol in T1-t2 (10.0%). Bornyl acetate has been found as a major constituent of *A. absinthium* oil from Cuba [22]. (*Z*)-2,6-dimethylocta-5,7- dien-2,3-diol (4) was found in T1-e4 and T1-e5 (the last two crops from Ejea) in relatively large amounts (30 and 54.6%) (Table **2**) [27].

(*Z*)-2,6-Dimethylocta-5,7-dien-2,3-diol (4) has been isolated from an acetone extract of *A. absinthium* T1-e2 and identified by NMR (Fig. **1**) [27]. Compound 4 has been reported in *Passiflora quadrangularis* [33], in the glandular haired accessions of *Medicago* spp. resistant to the alfalfa weevil [34] and in the aerial parts of *Aster bakeranus* [35].

The variations observed in *A. absinthium* EO composition were mostly quantitative but also qualitative, with the samples grown in controlled environments (SN) showing a more simplified composition than the pre-domesticated T1 ones (crops e and t) except for T1-e4,5. Moreover, the wild collected sample V showed fewer components than the cultivated ones. Chemical variations have been described for *A. absinthium* essential oils from plants cultivated *in vitro* (absence of monoterpenes and thujones and higher amounts of heavier compounds), in the greenhouse and in the field (presence of high amounts of monoterpenes and thujones) [36].

Two chemotypes have been described from the Iberian Peninsula [26]; *cis*-epoxyocimene type (with more than 50% of this compound) which was predominant in all the populations and a *cis*-epoxyocimene + chrysanthenyl acetate type (with 25-65% of *cis*-epoxyocimene and 15-50% of chrysanthenyl acetate). T1 (e and t), and V oils fitted the *cis*-epoxyocimene + chrysanthenyl acetate type; SNA the *cis*-epoxyocimene type, while the other SN oils did not follow that classification. Furthermore, if linalool oxide derived from *cis*-epoxyocimene as suggested [37], SNI could be included in the first group. Thujone was often found to be the major component of *A. absinthium* oil but none of the plants analyzed contained thujone. Wormwood essential oils from other locations had *cis*-chrysanthenol as the main component [13, 16], *cis*-epoxyociene and β-thujone / *cis*-chrysanthenyl acetate, isoascaridol / *trans*-sabinyl acetate among other compositions [25, 37].

Domesticated Wormwood Essential Oil

The domesticated T2 population originated from seeds collected from thujone-free plants submitted to a previous domestication period of six years (Pre-domesticated Teruel population T1) and individually selected based on their biomass and HD essential oil yield at the end of the cycle [28, 29].

The pre-domesticated SN1 population originated from wild germplasm (Sierra Nevada population SN0). The seeds were collected from individuals grown in a greenhouse and originated from a population donated by the Sierra Nevada National Park (Granada) (SN0) [29].

The seedlings were planted in 2008 in Ejea de los Caballeros, Zaragoza, Spain. Along with the cultivated plants, 40 individuals from each parent population (T1 and SN0) were maintained in a separated plot (1 row / parent population) of the same field for comparison purposes. Briefly, 6 randomized blocks with 6 rows / block and 30 individuals / row were planted for each population. Flowering plants were harvested yearly between 2008 and 2013 and processed for agronomic evaluation and extraction. For hydrodistilled (HD) essential oils, each block was extracted individually by pooling 20 plants collected from the central rows (3 and 4, 10 plants from each row) and only block 2 (BII) oils were used for chemical analysis and bioassays. For semi-industrial vapor pressure extraction (VP), plants harvested from all the blocks of each population were pooled (1,080 plants / population) and extracted [32].

Table **3** shows the average (range) composition of the hydro-distilled extracts (HD) from three years (2008-2010), analyzed individually. Qualitative and quantitative variations were observed between the two populations. Among the most variable compounds were cis-epoxyocimene (1), with higher variation for the T population; chrysanthenyl acetate (3); *(5Z)*-2,6-dimethylocta-5,7-dien--2,3-diol (4) with higher levels in the T samples. Camphor (6) and 1.8-cineol correlated with the domestication level of the population. Among the less variable compounds were (-)-*cis*-chrysanthenol (2), with higher levels in the SN population; *trans*-caryophyllene; linalool and an unknown sesquiterpene alcohol (Fig. **2**) [32].

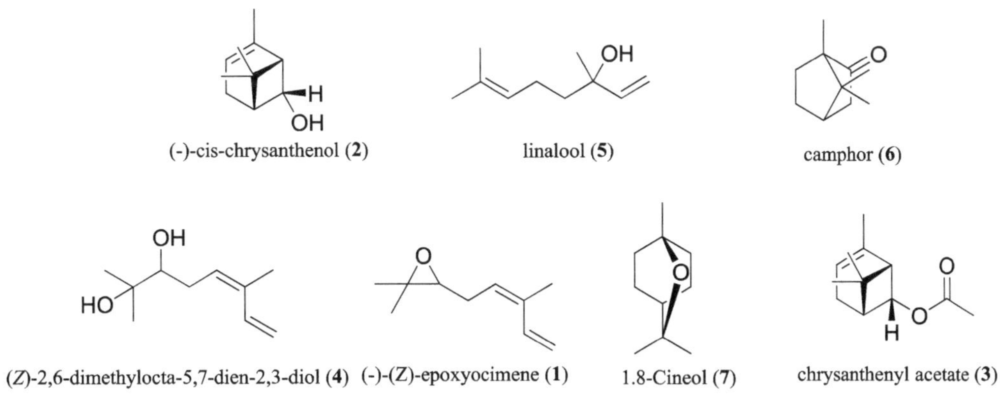

Fig. (2). Chemical structures of compounds **1-7**.

Table **4** shows the average (range) composition of the semi-industrial VP extracts from six years (2008-2013), analyzed individually. Similarly, qualitative and quantitative differences were also observed. As previously shown for the HDs, the presence of 1.8-cineol and camphor correlated with the domestication level of the population. (-)-*cis*-Chrysanthenol (2) showed 1.5 times higher levels in the SN1 population while chrysanthenyl acetate (3) was 22 times higher in T2. (*Z*)-2,--dimethylocta-5,7-diene-2,3-diol (4) was absent in all the VP extracts probably due to the extraction method [32].

Table 4. Chemical composition of essential oils (semi-industrial vapor pressure extracts, VP) of the different *A. absinthium* samples (abundance: % area). Values are expressed as average (range) of individual samples from six years (2008-2013). T2, domesticated Teruel population; SN1, pre-domesticated Sierra Nevada population.

	VP (2008-2013)	
Compounds	**T2**	**SN1**
α-pinene	2.1 (1.5 - 2.7)	1.4 (1.0 - 1.7)
1.8-cineol	2.1 (1.7 - 2.4)	0.1 (0 - 0.3)
linalool (**5**)	3.8 (2.1 - 5.1)	3.2 (1.5 - 4.0)
(-)-(*Z*)-epoxyocimene (**1**)	43.0 (38.9 - 44.1)	44.6 (37.3 - 49.0)
(*E*)-epoxyocimene	1.4 (0 - 3.0)	2.3 (1.8 - 3.1)
camphor (**6**)	4.7 (2.8 - 7.7)	0.2 (0 - 1.1)
(-)-*cis*-chrysanthenol (**2**)	11.9 (10.4 - 12.7)	18.1 (15.9 - 21.9)
chrysanthenyl acetate (**3**)	6.6 (5.5 - 7.4)	0.3 (0 - 0.7)
trans-caryophyllene	4.0 (3.4 - 4.2)	5.2 (4.3 - 5.6)
γ-cadinene	1.8 (1.3 - 2.8)	2.8 (1.7 - 4.1)

Overall, there was lower chemical variation for the VPs than for the HDs due to the sample size (pooled plants from three experimental blocks, six years analyzed) and the extraction method. A dendrogram based on the composition of the HDs (Fig. **3**) showed five groups: T2(08-10), T1(09)-SN1(09), T1(10)-SN1(08-1-)-SN0(08-10), SN0(09) and T1(10). These groups indicated a larger variation for the parent (T1 and SN0) and the pre-domesticated SN1 populations than the domesticated T2 [32].

A dendrogram based on the composition of the VPs (Fig. **4**) showed two major groups: T2 (except T210) with high levels of (*Z*)-2,6-dimethylocta-5,7-dien--2,3-diol (4) probably due to an uncontrolled variation in the semi-industrial extraction process, data not shown) and SN1, with higher variation for this second cluster. Therefore, the domestication of the T2 population resulted in lower chemical variation of the essential oils (HDs and VPs), allowing for the registration of a new *A. absinthium* variety (®Candial) [32].

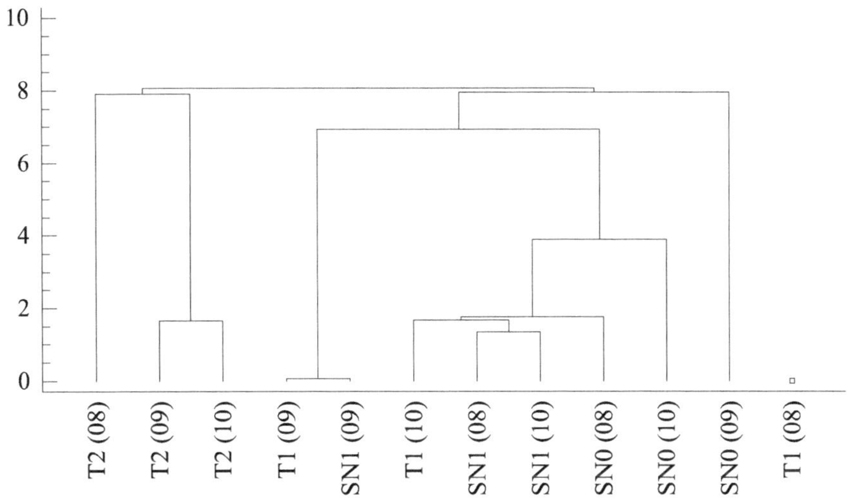

Fig. (3). Dendrogram (Nearest Neighbor Method,Squared Euclidean) generated from cluster analysis of GC-MS data of hydrodistilled extracts (HD) from two populations of cultivated *A. absinthium.* T, Teruel population; SN, Sierra Nevada population; T2, domesticated T population; T1, parent T population; SN1, pre-domesticated SN population; SN0, parent SN population; in parenthesis the crop date (year).

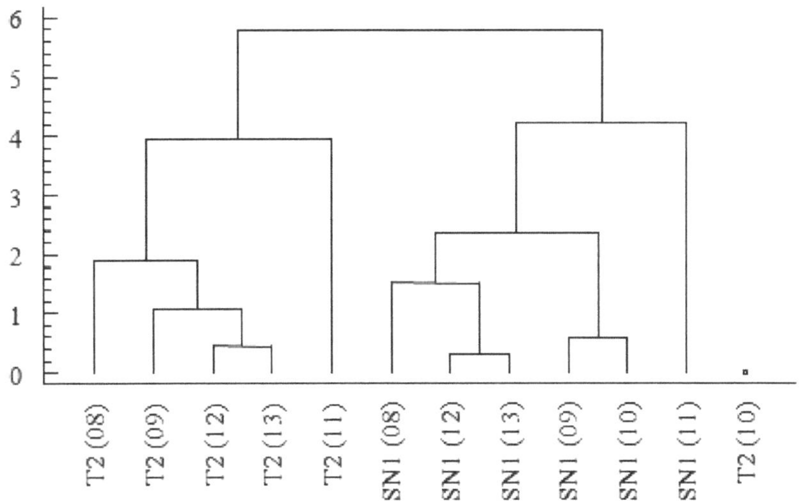

Fig. (4). Dendrogram (Nearest Neighbor Method,Squared Euclidean) generated from cluster analysis of GC-MS data of semi-industrial vapor-pressure extracts (VP) from two populations of cultivated *A. absinthium.* T2, domesticated Teruel population; SN1, pre-domesticated Sierra Nevada population; in parenthesis the crop date (year).

The HDs of the domesticated T2 population showed *cis*-epoxyocimene + *cis*-chrysanthenol chemotype, similar to the T1 parent population [27] but also contained chrysanthenyl acetate. However, the SN1 population showed a cis-epoxyocimene + chrysanthenyl acetate type, suggesting a high phenotypic plasticity for the production of these compounds [32].

Extraction Optimization

Supercritical Fluid Extraction (SFE) is an advanced separation method that complies with the principles of Green Chemistry and has advantages when extracting bioactive compounds from a natural matrix [38]. Particularly, the use of CO2 as a supercritical (SC) solvent, allows for mild extraction conditions due to the low critical temperature (31 ºC) and pressure (7.1 MPa) of the solvent, thus circumventing the thermal degradation of the extracted compounds [39]. Supercritical extraction of wormwood with carbon dioxide has been previously applied for the quantitative removal of thujone [40]. Furthermore, a pre-domesticated spanish *A. absinthium* population (T1e4) has been submitted to SFE using different extraction pressures and temperatures [24]. Approximately 90 g of grounded material, obtained from aerial parts, was extracted. An extra SFE experiment was performed with previously extracted plant material (18.0 MPa at 40 ºC) with the addition of 50 ml of EtOH as a co-solvent [24]. The experimental conditions of the selected experiments are shown in Table **5**.

Table 5. Supercritical experiments, plant material used and extraction conditions.

Experiment	Origin	Crop	P (MPa)	T (ºC)	ρ_{sc} (kg/m^3)	Flow rate (kg/h)
SCA	Teruel	T1e4	9.0	40	485.5	1.08
SCB		T1e4	13.5	40	753.6	1.08
SCC		T1e4	18.0	40	819.5	1.08
SCD		T1e4 + EtOH	18.0	40	819.5	0.81

Table **6** shows the compounds identified from the dichloromethane soluble fraction of different extracts and Table **7** shows their composition. The composition of the HD extracts was included for comparative purposes. The evolution of the composition of the SC extracts during the extraction process showed that different combinations of pressure and temperature affected the extract composition. For the 2005 crop, the increase in pressure (from SCA to SCC) resulted in decreased extraction levels for monoterpenes (up to compound 8) and an increased extraction of the sesquiterpene hydroxypelenolide isomer (16) and the heavier compounds (17-21). The addition of EtOH (SCD) resulted in a more efficient extraction of heavier compounds (16-21). The head fractions, H (collected from separator 2 at the initial stages of the extraction) and the tail fractions, T (collected from separator 2 at the final stages from the extraction) were richer in volatile compounds than the so-called waxes, W fractions (collected in separator 1 along the whole extraction).

Table 6. Compounds identified by GC-MS in the extracts.

+M	Base peak	Main fragments	IK	Compound	Number	Group
116	43	59-58-101	815	4-hydroxy-4-methyl-2-pentanone[1]	**7**	A lower terpenes
136	71	93-41-43-55-69	1083	linalool[1]	**5**	
152	79	81-77-41-39-53	1111	(*Z*)-epoxyocimene[1]	**1**	
152	81	91-109-41-79-39	1144	chrysanthenol[1]	**2**	
170	43	55-59-71-84-82	1220	dimethyl-octadienol $C_{10}H_{18}O_2$ isomer[2]	**4**	
170	43	55-71-59-84-81	1232	dimethyl-octadienol $C_{10}H_{18}O_2$ isomer[2]	**4**	
170	43	119-43-71-91-109	1263	dimethyl-octadienol $C_{10}H_{18}O_2$ isomer[2]	**4**	
170	43	55-74-59-84-81	1266	$C_{10}H_{18}O_2$ isomer	**8**	
194	59	43-81-67-82-79	1242	*trans*-chrysanthenyl acetate[1]	**3**	
204	93	133-91-79-69-41	1409	*trans*-caryophyllene[1]	**9**	B sesquiterpenes
204	93	41-79-91-105-107	1473	*β*-selinene[1]	**10**	
220	41	79-93-43-69-91	1559	caryophyllene oxide[1]	**11**	
?	69	41-57-93-68-121	1579	unidentified	**12**	
222	43	41-161-67-81-204	1630	sesquiterpene alcohol $C_{10}H_{18}O$[3]	**13**	
?	96	41-55-43-81-107	1638	unidentified	**14**	
248	215	43-159-230-119-91-248	1997	absilactone type sesquiterpene[2]	**15**	
250	232	43-217-162-150-77	1927	hydroxypelenolide isomer[2]	**16**	
268	43	58-71-41-57-59	1828	6,10,14-trimethyl-2-pentadecanone[1]	**17**	C ketones and diterpenes
284	119	132-145-41-69-105	1934	diterpene $C_{20}H_{28}O$[3]	**18**	
284	119	132-145-105-41-123	1986	diterpene $C_{20}H_{28}O$[3]	**19**	
286	119	121-93-105-41-91	1989	diterpene $C_{20}H_{30}O$[3]	**20**	
284	119	132-145-105-91-41	1993	diterpene $C_{20}H_{28}O$[3]	**21**	

[1] [41] [2] [27] [3] [26].

Table 7. Composition of the different extracts analyzed by GC-MS.

Compound	Abundance (% area)										
	Teruel (T1e5)										
	05HD	SCA		SCB			SCC		SCD		
		H	T	H	T	W	H_1	H_2	H	T	W
7	-	-	1.62	-	-	-	-		-	-	-
5	-	-	1.40	0.37	-	-	-		-	-	-
1	1.07	1.44	15.10	5.92	4.07	4.75	-		-	-	-
2	9.31	9.44	13.94	4.88	11.90	7.58	2.79	3.38	3.66	8.92	-
4	-	1.95	-	0.85	-	1.21	-		-	-	-
4	-	2.60	-	1.06	-	1.54	-		-	-	-
4	30.03	3.55	2.78	1.75	2.82	2.90	1.64	2.33	-	-	2.71
8	-	14.97	16.94	9.77	16.74	12.88	7.66	10.49	7.85	13.36	-
3	2.37	10.99	12.50	5.02	5.72	5.42	1.67		-	-	-
9	-	3.22	2.73	1.48	-	1.63	-		-	-	-
10	-	7.37	6.35	3.21	3.06	3.58	1.92	2.22	-	-	-
11	2.60	7.38	4.82	4.97	4.74	5.70	4.63	5.70	-	-	-
12	-	2.36	0.58	1.52	-	1.61	-		-	-	-
13	2.67	4.70	4.46	4.62	4.41	4.60	4.70	6.23	4.87	5.57	2.67
14	-	1.60	-	1.58	-	1.49	1.72		-	-	-
15	17.72	-	5.82	11.04	11.11	12.29	24.40	24.40	32.07	28.94	39.72
16	6.21	1.71	-	3.60	4.51	3.62	7.28	7.23	10.42	9.10	11.56
17	2.22	1.21	-	1.69	-	1.48	2.58	2.85	3.36	-	2.70
18	-	-	-	1.14	-	1.06	1.59	1.62	-	-	-
19	7.53	4.24	2.86	6.39	-	5.85	9.50	10.25	12.14	11.13	8.32
20	-	-	-	1.70	7.26	1.58	2.49	2.84	-	-	-
21	10.62	5.17	1.62	8.70	-	8.47	13.60	14.51	17.15	15.75	13.01

The SCA extracts were composed mostly of group A compounds (lower terpenes,) with (*Z*)-epoxyocimene (1), chrysanthenol (2, 8 and *trans*-chrysanthenyl acetate (3) being the major ones, their group B compounds (sesquiterpenes) being dominated by *β*-selinene (10) and caryophyllene oxide (11) followed by the sesquiterpene alcohol (13) and the sesquiterpene 15 (in fraction T) while their group C (ketones and diterpenes) had low amounts of diterpenes 19 and 21 in fraction H. The SCB extracts had a similar composition in group A compounds with 8 and 2 being the major ones; their group B compounds consisted of 15 (the major one), 11 and 13, 10, hydroxypelenolide 16 and small amounts of *trans*-caryophyllene (9) and their group C showed higher amounts of 19 and 21 with the presence of 20 in fraction T.

The SCC extracts were quite similar in composition since H1 and H2 are consecutive extraction fractions (T was not available), and they had a simpler composition in group A compounds with 8 being the major one; their group B composition had 15 as the major component followed by 16, 13 and 11, while their group C was similar to that of SCB-W with higher amounts of 19 and 21. The SCD extracts had a simpler group A (8 and 2) and B (15, 16 and 13) compositions than SCC with higher concentrations of 15, 16, 19 and 21 than the previous supercritical extracts.

Essential Oil Valorization

The global crop protection chemicals market is projected to reach a value of USD 70.57 billion by 2021, growing at a CAGR of 5.15% from 2016 to 2021. The market is driven by factors such as rising crop loss due to pest attacks, growing need for food security for the increasing population, and rising preference for biopesticides. The changing farming practices from traditional to conventional clubbed with advanced technologies and promotion of Integrated Pest Management (IPM) by the governments of many countries across the world drives the demand for crop protection chemicals in the agriculture sector.

On the basis of type, crop protection chemicals are segmented into herbicides, insecticides, fungicides and others which include nematicides. Among these, herbicides accounted for the largest market share in 2015, followed by insecticides and fungicides (http://www.researchandmarkets.com/publication/mxnuq69/3803208)*Artemisia* species have traditionally been used as anthelmintics [40] and different extracts and components (such as artemisinin from *A. annua*) have been reported as plant [42] and animal [43] nematocidal agents. *Artemisia absinthium* (wormwood) is an aromatic and medicinal plant of ethnopharmacological interest [1] with reported antihelminthic effects [44]. In addition, thujone-rich oils have been shown to have acaricidal [7], insecticidal [8 - 10] and fungicidal effects [10] and myrtenol-rich oils repelled fleas, flies, mosquitoes [11] and ticks [12].

In this context, the main objective of the domestication of thujone-free A. absinthium was to provide traceable plant biomass with a stable chemical composition to be used as a biopesticide for crop protection (see chapter I.6). Additional valorization of the domesticated wormwood (®Candial) essential oil against parasites (parasitic nematodes and protozoal parasites) has been performed given its traditional medicinal use (see chapter I.3).

The selected insect targets included important crop pests with different trophic adaptations (polyphagous / olyphagous): two chewing species (the lepidoteran *Spodoptera littoralis* and the beetle *Leptinotarsa decemlineata*) and two sucking species (the aphids *Myzus persicae* and *Rhopalosiphum padi*). The selected fungal plant pathogens included three species of *Fusarium* and *Botrytis cinerea*, all of them of great commercial importance. *Meloidogyne javanica* was the selected plant parasitic nematode specie. Root-knot nematodes (*Meloidogyne* spp.) are major threats to agriculture worldwide [45] and, in the last decades, environmental and human health concerns have steadily reduced the availability of efficient commercial nematicides. The selected target animal parasites were: the nematode *Trichinella spiralis*, responsible for the disease trichinellosis with an extensive geographical distribution as well as a wide range of hosts [46]. Trichinellosis presents an estimated global prevalence of 11 million of infected people [47]. Benzimidazoles, are broad spectrum drugs used against trichinellosis, however these compounds have low bioavailability [48] and resistance against them has been reported in various target nematodes [49]. The parasitic protozoa *Trypanosoma cruzi* are the aethiologic agent of Chagas disease, a frequently fatal illness affecting the heart and gastrointestinal systems. An estimated 8 million people in Latin America are infected with this pathogen and it is also spreading to the USA, Canada, and many parts of Europe and the Western Pacific as a result of migratory flows [50]. Only two drugs, nifurtimox and benznidazole, are in use against chronic infections and both have limitations, due to the need of a large number of doses over a long time period, side effects, and lack of effectiveness against all stages of the disease and all strains of the parasite. *Trichomonas vaginalis*, is the major cause of vaginitis, cervicitis and urethritis in women, and may cause nongonococcal urethritis, prostatitis and other genitourinary tract syndromes in men. Trichomoniasis is among the world´s most common sexually transmitted diseases with an annual incidence of more than 276 million cases per year [51]. A single drug, metronidazole, is currently available for treating trichomoniasis. However, metronidazole resistant strains have been found in unsuccessfully treated patients and some adverse effects have been described [52].

Insect Antifeedant Effects

The antifeedant effects of the T1pre-domesticated population and artificially cultivated *A. absinthium* essential oils against *Rhopalosiphum padi*, *Myzus persicae* and *Spodoptera littoralis* have been reported (Table **8**) [27].

Table 8. Antifeedant effects of *A. absinthium* essential oils from pre-domesticated (T1, crops e-t) and wild (artificially cultivated SN and field collected V) samples (100 μg/cm²).

Sample	*R. padi* (%SI[a])	*M. persicae* (%SI[a])	*S. littoralis* (%FI[b])
T1e1	26± 9	40 ± 8	34 ± 15
T1e2	67* ± 5	55* ± 9	81* ± 10
T1e3	26± 7	61* ± 8	50 ± 21
T1e4	31 ± 7	16 ± 6	21 ± 9
T1e5	40± 7	34 ± 6	21 ± 13
T1t2	25 ± 6	22 ± 7	39 ±17
T1t3	66* ± 7	49 ± 7	49 ± 20
T1t4	30 ± 9	42 ± 8	58 ± 19
T1t5	34 ± 6	26 ± 8	48 ± 22
SNC	39 ± 8	43 ± 8	48 ± 10
SNI	25 ± 7	26 ± 8	54 ± 20
SNA	55± 6	32 ± 9	70± 6
V	39 ± 6	42 ± 6	26 ± 17

[a]Percent settling inhibition; [b]Percent feeding inhibition.*$p<0.05$, Wilcoxon Paired Rank Test.

The samples rich in *cis*-epoxyocimene (1) (>45%) (T1-e2, t4 and SNA) were the most active ones against *S. littoralis*. Moderate-low effects (%SI>60) were found for T1-e2, t3 and e3 against the aphids [27].

The result of a step-wise multiple regression performed between the oil composition of samples T1e, t and V and their antifeedant effects indicated that the activity against *M. persicae* could be partially explained by linalool (5) (35.8%, R2 adjusted, p=0.031), and against *S. littoralis* by *cis*-epoxyocimene (1) (36.8%, R2 adjusted, p=0.028). There was not significant relationship between any of the components and the antifeedant effects against *R. padi*, suggesting synergistic effects [27]. Linalool has been reported as an antifeedant and repellent to *M. persicae* [53] while the (+)-(*R*)-enantiomer of (*E*)-6,7-epoxyocimene has been previously identified as a pheromone emitted by the male fruit-spotting bug *Amblypelta nitida* [54].

Furthermore, the study of the antifeedant effects of the domesticated (T2) and pre-domesticated (SN1) populations [32] showed that overall, the antifeedant effects of the *A. absinthium* essential oils (HD and VP) were moderate-low (Table **9**). The Colorado potato beetle, *Leptinotarsa decemlineata*, was affected by cultivated (T2-VP) and wild (SN0-HD) extracts. *S. littoralis* was affected by cultivated T2 and SN1 HD extracts. The aphids (*M. persicae* and *R. padi*) were not affected. Similarly, the antifeedant effects of HDs from the parent germplasm of both populations studied here (T1, SN0) showed no effects on *M. persicae* and low-moderate on *S. littoralis* [27].

Table 9. Antifeedant effects of two populations of cultivated *A. absinthium* (Teruel and Sierra Nevada) extracts (HD, hydrodistilled; VP, semi-industrial vapor-pressure extracts) on the target insect species. Values are expressed as average (range) of individual samples from three years (2008-2010) tested at 100μg/cm².

Germplasm	Extract	Teruel				Sierra Nevada			
		L. decem-lineata	*S. littoralis*	*M. persicae*	*R. padi*	*L. decem-lineata*	*S. littoralis*	*M. persicae*	*R. padi*
		%FI[a]		%SI[a]		%FI[a]		%SI[a]	
Parent	HD	58.4	52.9	49.1	47.3	82.1	47.8	53.7	49.1
		(57.5 - 59.7)	(40.1 - 64.6)	(33 - 72.9)	(45.4 - 50.9)	(75.9 - 86.6)	(40.8 - 53.2)	(33.2 - 66.6)	(38.1 - 58.3)
Cultivated	HD	56.9	66.3	40.8	30.0	52.5	69.3	39.2	32.8
		(54.7 - 59.5)	(54.9 - 74.2)	(25.7 - 50.4)	(27.7 - 32.4)	(47 - 62.3)	(65.9 - 71.1)	(37.5 - 40.9)	(25.5 - 46.7)
	VP	73.2	54.2	51.7	18.4	43.7	56.3	45.6	21.2
		(63.5 - 86.3)	(43.3 - 62.8)	(47.6 - 54.2)	(8.4 - 23.6)	(37.2 - 52.2)	(45.9 - 65.1)	(38.3 - 51.7)	(8.5 - 37.7)

[a]Percent feeding (FI) / setting (SI) inhibition.

Other wormwood EOs and their major components (camphor, 1.8-cineol, terpinen-4-ol, borneol, bornyl acetate and α-terpineol) have reported insecticidal effects against the granary weevil *Sitophilus granarius* (Coleoptera) [9]. Thujone rich *A. absinthium* oils from Uruguay had insecticidal effects against *Tuta absoluta* and *Trialeuroides vaporariorum* [10].

Almost all of the supercritical extracts exhibited an antifeedant activity of higher than 80% (Table **10**). A similar pattern was found for the effects of these extracts on *M. persicae* and *R. padi*, aphids being less sensitive than *S. littoralis*. The SCA and SCC extracts were the most active against both aphids. Overall, the H and T fractions were more active than the W one for experiments SCC-SCF, while experiments SCA and SCB did not follow this pattern. The antifeedant effects of the SC extracts did not correlate with the presence of any given compound. Therefore a synergistic effect of these mixtures was suggested [24, 30].

Table 10. Antifeedant effects of the different *A. absinthium* extracts on the target insect species

Extract	*S. littoralis* (100μg/cm²)	*M. persicae* (100μg/cm²)			*R. padi* (100μg/cm²)		
	FI%[b]	SI%[c]	%C[d]	%T[ew]	SI%[c]	%C[d]	%T[e]
HD05[a]	19.8±16.7	30.8±6.8	53±5	47±5	34.7±7.9	40±5	60±5
OSE05[a]	21.3±9.8	16.5±6.3	54±5	46±5	30.7±7.6	59±4	41±4
SCA-H	82.8±4.7*	79.5±6.1*	84±4	16±4	81.7±4.0*	86±3	14±3
SCA-T	93.7±4.0*	63.4±7.2*	75±4	25±4	90.6±2.0*	92±2	8±2
SCA-W	92.0±7.2*	48.1±11.6	67±7	33±7	29.8±6.0	58±3	42±3
SCB-H	93.5±3.7*	46.0±6.7	66±4	34±4	55.4±6.0	70±3	30±3
SCB-T	85.6±10.0*	59.6±7.7*	75±4	25±4	40.2±7.2	61±4	39±4
SCB-W	86.4±8.4*	-	-	-	28.1±7.6	55±5	45±5
SCC-H1	88.6±4.9*	96.7±3.3*	97±3	3±3	83.1±6.4*	87±5	13±5
SCC-H2	64.7±9.3	94.3±2.2	95±2	5±2	58.9±7.1	73±4	27±4
SCD-H	90.9±8.4*	83.1±4.3*	87±3	13±3	68.8±5.8*	78±3	22±3
SCD-T	85.7±8.7*	78.5±5.2*	84±3	16±3	68.0±4.9*	77±2	23±2
SCD-W	84.2±9.3*	72.4±3.7*	80±2	20±2	34.8±6.8	61±3	39±3

[a] [27] [b]FI%: feeding inhibition index: [1-(treatment consumption/control consumption)] x 100 [b]SI%: settling inhibition index: [1-(treatment/control)] x 100 [d]% C: Percentage of individuals remaining on the control surface. [e]% T: Percentage of individuals remaining on the treated surface. SC: supercritical extract, conditions shown in Table **1**. H refers to head extract, T to tail extract, W to waxes. HD: hydrodistillation OSE: organic solvent extract A: aeroponic growth 2005, 2006, 2008: different crops *P<0.05, Wilcoxon Paired Rank Test

Antifungal Activity

T1pre-domesticated and artificially cultivated *A. absinthium* essential oils were also submitted to antifungal assays against three species of *Fusarium* [27]. *F. oxysporum* and *F. solani* were affected by T1-e1and t2, with significant amounts of cis-epoxyocimene (Table **11**).

Table 11. Antifungal effects of *A. absinthium* essential oils from cultivated (T1 crops e-t) and wild (artificially cultivated SN, field collected V) samples against *Fusarium* spp. ED50 (μg/ml) (upper and lower 95% confidence limits).

Sample	*F. moniliforme*	*F.oxysporum*	*F. solani*
T1e1	>1	0.5 (0.4, 0.)	0.6 (0.2, 0.9)
T1e2	>1	>1	>1
T1e3	>1	>1	>1
T1e4	>1	0.8 (-0.1, 1.2)	>1

(Table 11) contd.....

Sample	*F. moniliforme*	*F.oxysporum*	*F. solani*
T1e5	>1	>1	>1
T1t2	0.8 (0.7, 0.9)	0.58 (0.5, 0.7)	0.6(0.5, 0.7)
T1t3	>1	>1	0.8 (0.5, 1.0)
T1t4	>1	0.9 (0.6, 1.2)	>1
T1t5	>1	0.9 (0.7, 1.2)	0.9 (0.7, 1.2)
SNC	>1	>1	>1
SNI	>1	0.8 (0.5, 1.0)	>1
V	>1	>1	>1

$\mathbf{ED_{50}}$: effective dose to give a 50% of inhibition.

Similarly, the essential oils from domesticated Spanish wormwood were tested against these pathogens [32]. Table **12** shows the average (range) antifungal effects of essential oil extracts from three years (2008-2010) tested individually. The semi-industrial vapor pressure extracts (VP) showed stronger antifungal effects that the HDs against all the *Fusarium* species, with the SN1 VP extracts being more effective against *F. moniliforme* and *F. solani* than the T2 ones. These extracts were further tested against *Botrytis cinerea,* being this species the most sensitive one. The antifungal effects of HDs from both parent populations (T1, SN0) against *Fusarium* spp. were lower than the ones found for the VPs. Therefore, for antifungal applications the domesticated population submitted to vapor-pressure extraction should be used [27].

Table 12. Antifungal effects (mycelial growth inhibition) of two populations of cultivated *A. absinthium* (Teruel and Sierra Nevada) extracts (HD, hydrodistilled; VP, semi-industrial vapor-pressure) on the target fungal species. Values are expressed as average (range) individual samples from three years (2008-2010) tested at 1mg/ml.

Germplasm	Extract	Teruel				Sierra Nevada			
		F. moniliforme	***F. oxysporum***	***F. solani***	***B. cinerea***	***F. moniliforme***	***F. oxysporum***	***F. solani***	***B. cinerea***
Parent	HD	31.0	42.2	34.0	nt	34.2	45.0	59.7	nt
		(25.1 – 37.0)	(38.0 - 46.5)	(19.2 - 48.8)	nt	(19.9 - 48.4)	(27.7 - 62.0)	(49.8 - 69.5)	nt

(Table 12) contd.....

Germplasm	Extract	Teruel				Sierra Nevada			
		F. moniliforme	*F. oxysporum*	*F. solani*	*B. cinerea*	*F. moniliforme*	*F. oxysporum*	*F. solani*	*B. cinerea*
Cultivated	HD	36.7	51.1	54.2	nt	36.8	56.8	42.2	nt
		(28.1- 45.2)	(45.5 - 56.7)	(37.2 - 71.1)	nt	(31.1 - 42.5)	(48.9 - 64.6)	(19.6 - 64.8)	nt
	VP	100	100	100	100	100	100	100	100
	ED_{50} [a]	(0.24 - 0.43)	(0.29 - 0.40)	(0.24 - 0.50)	(0.01 - 0.07)	(0.11 - 0.24)	(0.20 - 0.50)	(0.11 - 0.27)	(0.01 - 0.03)

[a] ED_{50}, concentration needed to produce 50% mycelial growth inhibition (mg/ml)

A. *absinthium* essential oils containing (*Z*)-epoxyocimene and chrysanthenyl acetate as major components inhibited the growth of *Candida albicans* and Saccharomyces cerevisiae var. chevalieri [5]. Wormwood essential oil from a Turkish population, characterized by camphor, 1.8-cineol and chamazulene, was fungicidal against 34 species including *F. solani* and *F. oxysporum* [9]. *A. absinthium* EO from Uruguay rich in thujone showed antifungal effects against *Alternaria* sp. and *Botrytis cinerea* [10]. However, none of these studies identified the compound/s responsible for these antifungal effects.

Table 13. Antifungal effects (mycelial growth inhibition) of fractions 1-9 from T2 (11) VP extract and compounds 1-3 on the target fungal species.

Sample		Concentration (mg/ml)	*F. oxysporum*	*B. cinerea*
Fraction	1	1.0	31.19 ± 2.6	46.62 ± 1.4
	3	1.0	54.62 ± 2.1	46 ± 1.0
	4	1.0	49.12 ± 3.1	47.03 ± 1.9
	5	1.0	51.07 ± 1.8	41.09 ± 1.4
	6	1.0	45.09 ± 4.2	35.13 ± 1.2
	7	1.0	100 ± 0.0	100 ± 0.0
		0.5	68.00 ± 1.8	45.37 ± 2.2
		0.1	-34.54 ± 5.5	-19.46 ± 2.3
	9	1.0	52.58 ± 4.5	29.28 ± 2.0

(Table 13) contd.....

Sample		Concentration (mg/ml)	*F. oxysporum*	*B. cinerea*
Compound	(-)-*cis*-chrysanthenol	1.0	100 ± 0.0	100 ± 0.0
		0.5	68.3 ± 1.5	na
		0.1	na	na
	linalool	1.0	70.38±2.1	100±0.0
		0.5	45.54±2.5	na
		0.1	7.58 ± 3.7	na
	camphor	1.0	7.6 ± 2.3	21.6±3.7

na, not available

The fractionation of a VP extract from the domesticated T2 population resulted in the identification of two antifungal fractions (7 and 8, Table **13**). Fraction 7 was rich in (-)-*cis*-chysanthenol (2), camphor (6) and linalool (5). Fraction 8 was also rich in (-)-*cis*-chrysanthenol (2) followed by linalool (5) and hexenyl butyrate (Table **14**). Further compound isolation and bioassay demonstrated that (-)-*cis*-chrysanthenol (2) was the main antifungal agent of the VP extracts, followed by linalool (5), present in lower amounts [32].

Table 14. Chemical composition of T2(11) VP extract and fractions 1-9 (abundance: % area).

Compound	VP	Fraction							
	T2(11)	**1**	**3**	**4**	**5**	**6**	**7**	**8**	**9**
Yield (%)	-	12.6	0.8	7.6	5.3	13.5	4.8	7.8	1.7
linalool (**5**)	2.0	-	-	-	-	-	-	17.3	-
(-)-(*Z*)-epoxyocimene (**1**)	39.8	-	-	13.5	63.9	86.6	5.6	-	-
camphor (**6**)	4.5	-	-	-	1.4	7.2	27.0	-	-
(-)-*cis*-chrysanthenol (**2**)	11.9	-	-	-	-	-	60.8	63.0	0.7
(*E*)-3-hexenyl butyrate	1.1	-	-	-	-	-	1.7	12.4	-
chrysanthenyl acetate (**3**)	5.3	-	1.0	35.7	22.1	0.9	-	-	-
(5*Z*)-2,6-dimethylocta-5,7-diene-2,3-diol (**4**)	2.0	-	-	-	-	-	-	-	67.8
trans-caryophyllene	3.8	29.5	-	-	-	-	-	-	-
germacrene-D	2.3	15.5	-	-	-	-	-	-	-
β-selinene	1.1	8.8	-	-	-	-	-	-	-
dihydrochamazulene	5.8	2.5	7.9	-	-	-	-	-	-
chamazulene	2.6	1.1	81.7	-	-	-	-	-	-

(-)-*cis*-chrysanthenol has not been previously reported as being antifungal. Linalool is a known antimicrobial and antifungal agent involved in plat defense against pathogens [55 - 57]. (*1R*)-(+)-camphor inhibited the mycelial growth of Choanephora cucurbitarum, a plant pathogen [55] but resulted inactive against the fungal species tested here [32].

Nematicidal Effects

The *in vitro* and *in vivo* nematocidal activity of VP-EOs from two populations (T2 and SN1) and crops (2009 and 2010, see Table **15** for detailed composition) of cultivated Spanish *A. absinthium* against two parasitic nematode models (the animal parasite *Trichinella spiralis* and the root-knot nematode *Meloidogyne javanica*) has been recently reported [58]. In this context, animal parasitic nematodes [59 - 62] and plant parasitic nematodes [63 - 66] have been increasingly targeted by studies on the effects of essential oils. However, most research on animal and plant parasitic nematodes takes place separately due to their differences in host species [67].

Table 15. ***Arthemisia. absinthium*** **EOs composition of two cultivated populations (T2 and SN1) collected in 2009 and 2010.**

Compound	Percentage abundance			
	T2		SN1	
	2009	2010	2009	2010
α-pinene	2.70			1.62
camphene	1.07			
β-myrcene	0.68		1.15	1.13
β-ocimene			1.54	1.27
1.8-cineol	2.43			
linalool (**5**)	4.28	3.42	3.49	3.54
[M]+152 (**81** 68-79-41-53)	1.25	0.54	1.94	0.49
cis-epoxyocimene (**1**)	38.92	25.89	42.72	38.77
trans-epoxyocimene	7.74	0.99	2.45	2.16
chrysanthenol (**2**)	12.71	30.32	23.48	22.49
(*E*)-3-hexenyl butyrate	1.26	1.34	1.30	1.64
chrysanthenyl acetate (**3**)	7.35		0.71	0.46
(*5Z*)-2,6-dimethylocta-5,7-diene-2,3-diol (**4**)	1.72	20.21	1.72	1.90
β-elemene		0.98	1.46	2.57
trans-caryophyllene	4.20	2.47		5.43

(Table 15) contd.....

Compound	Percentage abundance			
	T2		SN1	
germacrene D	1.36	1.15	5.99	1.86
sesquiterpene alcohol	0.51	1.79		0.74
7-Ethyl-1,4-dimethylazulene	1.09	0.87		

Table **16** shows the *ex vivo* activity obtained for the different EO concentrations tested against T. spiralis. All the EOs showed a high nematocidal activity. These oils were also active *in vivo* (larval reduction values > 50%) [58].

Table 16. ***Ex vivo*** **activity of different** ***A. absinthium*** **EOs (T2 and SN1 populations, 2009 and 2010) against** ***T. spiralis*** **L1 muscle larvae. Results are expressed as mean ± standard error (SE) of adult worms (AW) recovered from intestine of mice infected with** ***ex vivo*** **treated larvae (2 replicates).**

EO (mg/ml)	T2-09		T2-10		SN1-09		SN1-10	
	AW	LRa %*	AW	LRa %	AW	LRa %	AW	LRa %
0.03	97.13±9.96	13.18	100.99±7.4	9.83	81.63±7.87	27.04	81.23±6.37	27.37
0.06	90.88±6.38	18.77	80.13±7.72	28.38	74.38±6.65	33.52	76.38±8.96	31.73
0.12	71.00±11.48	36.54	60.50±4.87	45.92	67.75±8.31	39.44	65.00±7.88	41.90
0.25	58.88±8.32	47.37	57.00±6.05	49.05	46.13±10.08	58.77	59.00±11.88	47.26
0.50	27.13±4.58	75.75	31.00±6.29	72.29	18.88±5.84	83.13	19.13±6.69	82.91
1	1.64±0.71	98.54	0.09±0.36	99.92	0.73±0.11	99.35	0.00±0.00	100
CONTROL	111.9±8.35							

*Larval reduction rate ($p<0.05$)

Aqueous and ethanolic *A. absinthium* extracts reduced the number of *Haemonchus contortus* adult worms *in vitro* and the faecal egg output in sheep infected with *H. contortus, Trichuris ovis, Chabertia ovina, Bunostomum trigonocephalum* and *Oesophagostomum columbianum* [43]. An ethanolic extract of *A. absinthium* showed *in vitro* trematocidal effects against Schistosoma mansoni, Echinostoma caproni and Fasciola hepatica [68]. Furthermore, methanolic extracts of Artemisia, including *A. absinthium* inhibited trichinellosis in rats [69]. This is the first report on trichinellosis inhibition by *A. absinthium* essential oil.

The oils were not cytotoxic to macrophage cells (growth inhibition < 20%) at the highest concentration tested (200 μg/ml). Therefore, *cis*-epoxyocimene + cis-chrysanthenol chemotype *A. absinthium* EOs could go for further development as anthelmintic agents [58].

In contrast, these *A. absinthium* EOs were not active against the plant parasitic nematode *M. javanica* (Table **17**). Similarly, other *A. absinthium* EOs tested against the pine wood nematode *B. xylophilus* [70] and the stem nematode *D. dipsaci* [71] also gave negative results.

Table 17. Effects of *A. absinthium* EOs (T2 and SN1 populations, 2009 and 2010, 1 mg/ml) on mortality of second stage juveniles (J2) of *Meloidogyne javanica* (4 replicates).

EO	J2 mortality (%)* at 72h
T09	5.44 ± 1.83
T10	11.56 ± 2.57
SN09	8.51 ± 1.72
SN10	8.09 ± 0.43

*Mean values corrected according to Schneider–Orelli's formula

A. absinthium EOs were highly effective against animal (T. spiralis) but not against plant (*M. javanica*) endoparasitic parasitic nematodes, despite the similarities between them in some structures adapted to parasitism [72 - 75]. The different nematicidal effects of the tested EOs against both parasitic nematodes species could be related to the lack of close phylogenetic relationship and their different host-parasite interactions [67, 72].

Antiprotozoal Activity

The essential oils from pre-domesticated (T1) and artificially cultivated (SNC) plants showed moderate-high antiparasitic effects against *Leishmania infantum* and *Trypanosoma cruzi* (Table **18**) [27].

Table 18. Leishmanicidal and trypanocidal effects (% mortality) from *A. absinthium* essential oil from different populations.

Essential oil	***L. infantum***			***T. cruzi***		
μg/ml	**800**	**400**	**100**	**800**	**400**	**100**
T1e2	83 ± 0.2	88 ±0.6	91 ±0.6	94 ± 1.1	92 ±1.5	15 ±3.0
T1t2	79 ± 0.1	91 ±0.6	0	85 ±1.8	89 ±0.5	35 ±8.4
SNC	90 ±0.8	96 ± 0.3	80 ±0.9	95 ±0.6	99 ± 0.2	71 ±2.8

[a]Dose of 200 μg/ml; [b] Dose of 800 μg/ml.

The antiparasitic and cytotoxic effects of the vapor pressure essential oil (VP) from the domesticated *A. absinthium* T2 population (crop 2011) and its fractions were further tested against the protozoans *T. cruzi* and Trichomonas vaginalis. Additionally, their selective cytotoxicity was tested on several tumoral (A549, H292, HCT116, MCF7, SK-MEL5) and non tumoral (HS5) human cell lines [75].

The characterized VP oil (T2-2011) and its fractions (Table **19**) were tested (Tables **19** and **20**). The VP oil was active on *T. cruzi* (GI50 144.6 µg/mL) (Table **19**), in agreement with the bioactivity reported for the parent population essential oil [27]. This VP also showed trichomonacidal effect (Table **20**) (GI50 87.2 µg/mL). Among the VP fractions, VLC1 and 2 were the most active on *T. cruzi* (Table **19**). VLC1 showed the strongest trichomonacidal effect (Table **20**) [76].

Table 19. Trypanocidal activity of *A. absinthium* vapor pressure essential oil (VP T2-2011), fractions VLC1 to VLC9, *trans*-caryophyllene and (-)-*cis*-chrysanthenol. Nifurtimox is included as drug reference. Data are expressed as percentage of growth inhibition ± standard deviation.

VP, fractions and compounds	Concentration (µg/mL)					
	800	400	200	100	10	1
VP T2-11	100 ± 1.7	96.0 ± 11.2	83.6 ± 18.0	33.0 ± 7.2	5.8 ± 5.2	4.7 ± 4.9
VLC1	-	-	-	97.8 ± 4.6	0.6 ± 1.8	0.5 ± 1.1
VLC2	-	-	-	94.7 ± 8.4	1.5 ± 2.3	1.3 ± 2.2
VLC3	-	-	-	30.0 ± 5.4	0.9 ± 1.7	0.0 ± 0.0
VLC4	-	-	-	46.7 ± 15.3	2.0 ± 3.8	4.5 ± 4.2
VLC5	-	-	-	26.8 ± 15.9	1.0 ± 1.3	0.5 ± 1.8
VLC6	-	-	-	8.4 ± 6.4	1.6 ± 2.0	3.3 ± 4.3
VLC7	-	-	-	10.1 ± 7.6	0.0 ± 0.0	4.8 ± 5.9
VLC8	-	-	-	41.9 ± 10.1	11.1 ± 5.6	4.5 ± 3.7
VLC9	-	-	-	6.5 ± 8.1	0.8 ± 1.4	1.4 ± 2.3
trans-caryophyllene	-	-	-	99.4 ± 1,9	11.6 ± 14.1	5.4 ± 5.5
(-)-*cis*-chrysanthenol	-	-	-	3.5 ± 8.3	0.0 ± 0.0	0.4 ± 0.7
Nifurtimox	-	-	-	99.9 ± 0.2	100.0 ± 0.5	20.8 ± 2.8

Table 20. Trichomonacidal activity of *A. absinthium* vapor pressure essential oil (VP T2-11), fractions VLC1 to VLC9, *trans*-caryophyllene and (-)-*cis*-chrysanthenol. Data are expressed as percentage of growth inhibition ± standard deviation.

VP, fractions and compounds	Concentration (µg/mL)					
	500	250	100	75	37.5	18.75
VP T2-11	99.1 ± 0.4	87.4 ± 1.3	53.7 ± 0.5	51.7 ± 3.3	14.5 ± 2.4	0.8 ± 1.2
VLC1	99.3 ± 0.2	98.8 ± 0.2	81.6 ± 0.2	47.6 ± 3.8	10.3 ± 2.0	0.4 ± 0.5
VLC2	100 ± 0.0	44.3 ± 5.6	11.8 ± 4.0	2.7 ± 3.9	1.2 ± 1.7	0.0 ± 0.0

(Table 20) contd.....

VP, fractions and compounds	Concentration (µg/mL)					
	500	250	100	75	37.5	18.75
VLC3	35.5 ± 0.1	12.2 ± 2.9	10.3 ± 2.0	0.4 ± 0.5	1.0 ± 1.4	0.0 ± 0.0
VLC4	99.5 ± 0.1	99.4 ± 0.1	39.1 ± 0.2	27.0 ± 3.2	4.0 ± 0.1	0.0 ± 0.0
VLC5	99.3 ± 0.4	43.3 ± 2.6	2.4 ± 3.4	3.7 ± 5.3	0.0 ± 0.0	0.0 ± 0.0
VLC6	71.6 ± 1.3	36.5 ± 1.8	4.7 ± 6.6	2.5 ± 3.6	0.4 ± 0.6	0.0 ± 0.0
VLC7	100.0 ± 0.0	47.9 ± 4.3	27.0 ± 6.8	0.0 ± 0.0	0.0 ± 0.0	0.0 ± 0.0
VLC8	99.9 ± 0.6	92.2 ± 0.1	0.0 ± 0.0	0.0 ± 0.0	0.0 ± 0.0	0.0 ± 0.0
VLC9	4.7 ± 6.7	1.7 ± 2.5	0.0 ± 0.0	1.7 ± 2.4	0.0 ± 0.0	0.5 ± 0.7
trans-caryophyllene	100.0 ± 0.0	99.0 ± 0.2	80.3 ± 7.3	53.7 ± 2.2	34.3 ± 6.0	19.9 ± 8.5
(-)-*cis*-chrysanthenol	100.0 ± 0.0	96.3 ± 1.4	54.5 ± 5.9	45.4 ± 2.5	40.9 ± 0.8	2.1 ± 0.8

trans-Caryophyllene showed a remarkable activity against *T. cruzi* (GI50 39.2 µg/mL). This compound also showed significant trichomonacidal effect (Tables **19** and **20**) (GI50 68.7 µg/mL). The antiparasitic effects of VLC1 against *T. cruzi* and T. vaginalis could be partially attributed to *trans*-cariophyllene. The activity of VLC2 on *T. cruzi* cannot be attributed to chamazulene since fraction VLC3, containing 81.7% chamazulene, was inactive. Therefore, dihydrochamazulene or the mixture of both compounds could be responsible for this effect. (-)-*cis*-Chrysanthenol (2) showed activity against T. vaginalis (GI50 87.2 µg/mL) (Table **20**). The effects of VLC8 against T. vaginalis at high concentrations can be attributed to (-)-*cis*-chrysanthenol (2). Fractions VLC1 and 2 were not cytotoxic against non-tumor cell line HS5 (Table **21**), suggesting selective antiparasitic activity for these two fractions [75].

Table 21. Cytotoxic effects on A549, H292, HCT116, MCF7, SK-MEL-5 and HS5 cells of *A. absinthium* vapor pressure essential oil (VP T2-2011) and fractions VLC1-9. Paclitaxel and Cisplatin are included as drugs reference. Data are expressed as viability percentage ± standard deviation, relative to vehicle-treated control cells.

VP and fractions	Concentration	Cell lines					
		A549	H292	HCT116	MCF7	SKMEL-5	HS5
VP T2-11	250 µg/mL	79.5 ± 0.4	76.7 ± 1.1	51.1 ± 1.8	91.3 ± 0.9	60.1 ± 3.1	80.5 ± 3.4
	100 µg/mL	98.6 ± 5.2	82.5 ± 1.0	58.5 ± 3.5	95.7 ± 1.8	78.5 ± 0.9	86.7 ± 0.9
VLC-1	250 µg/mL	60.6 ± 1.5	73.0 ± 0.9	49.6 ± 1.3	95.7 ± 1.8	58.2 ± 0.8	80.8 ± 1.5
	100 µg/mL	74.8 ± 1.9	77.2 ± 3.2	52.1 ± 0.8	91.3 ± 0.9	56.9 ± 3.0	79.1 ± 2.4
VLC-2	250 µg/mL	84.0 ± 4.7	89.2 ± 7.8	59.4 ± 4.2	84.0 ± 9.8	58.8 ± 10.5	91.7 ± 6.8
	100 µg/mL	69.5 ± 10.2	90.4 ± 2.0	59.5 ± 3.6	89.6 ± 2.1	67.7 ± 5.2	83.3 ± 7.4

(Table 21) contd.....

VP and fractions	Concentration	Cell lines					
		A549	H292	HCT116	MCF7	SKMEL-5	HS5
VLC-3	250 µg/mL	83.6 ± 4.2	97.2 ± 7.1	59.0 ± 1.4	91.9 ± 3.5	57.3 ± 5.7	90.2 ± 5.2
	100 µg/mL	100 ± 9.1	98.1 ± 5.4	63.4 ± 0.9	93.3 ± 0.5	66.3 ± 5.5	84.1 ± 4.3
VLC-4	250 µg/mL	80.6 ± 2.5	73.5 ± 3.1	63.9 ± 1.3	88.7 ± 0.2	57.4 ± 1.0	78.5 ± 2.5
	100 µg/mL	95.9 ± 2.3	79.9 ± 0.8	78.5 ± 0.6	100 ± 0.7	58.8 ± 1.6	86.1 ± 1.8
VLC-5	250 µg/mL	57.7 ± 1.3	78.1 ± 0.6	51.7 ± 1.5	80.1 ± 3.7	54.1 ± 2.1	75.9 ± 1.7
	100 µg/mL	95.2 ± 3.1	79.7 ± 1.3	62.8 ± 3.6	97.3 ± 2.3	53.5 ± 3.9	68.2 ± 1.8
VLC-6	250 µg/mL	100 ± 2.5	87.8 ± 3.9	64.4 ± 2.2	96.9 ± 2.3	59.5 ± 4.2	81.2 ± 2.9
	100 µg/mL	100 ± 2.7	96.8 ± 1.8	75.0 ± 2.7	100 ± 1.1	72.0 ± 1.3	90.7 ± 3.7
VLC-7	250 µg/mL	100 ± 3.4	94.4 ± 1.1	93.1 ± 1.9	93.7 ± 1.6	69.5 ± 8.9	89.1 ± 1.8
	100 µg/mL	100 ± 2.3	100 ± 1.7	93.6 ± 2.3	97.9 ± 1.9	78.4 ± 5.1	95.7 ± 2.1
VLC-8	250 µg/mL	67.7 ± 4.6	82.8 ± 13.6	50.9 ± 2.0	69.1 ± 2.3	54.5 ± 1.9	76.3 ± 3.9
	100 µg/mL	100 ± 2.4	87.2 ± 5.2	65.0 ± 0.6	100 ± 1.7	54.5 ± 2.8	86.6 ± 2.8
VLC-9	250 µg/mL	100 ± 0.2	92.5 ± 6.0	96.8 ± 1.6	97.8 ± 1.7	65.9 ± 6.0	95.6 ± 1.7
	100 µg/mL	100 ± 3.1	97.4 ± 5.0	89.6 ± 3.5	96.1 ± 1.7	79.3 ± 2.1	95.4 ± 2.1
Paclitaxel	85 µg/mL	41.9 ± 2.9	41.8 ± 1.7	40.2 ± 2.8	52.9 ± 3.5	34.9 ± 2.4	27.7 ± 4.8
Cisplatin	300 µg/mL	34.6 ± 1.5	57.3 ± 2.6	25.7 ± 1.7	65.6 ± 1.1	42.6 ± 0.3	22.8 ± 2.4

When tested against tumor cell lines, the VP and fractions inhibited the growth of human tumor cell lines in a dose-dependent manner. The most sensitive cell lines were SK-MEL-5 (melanoma) and HCT116 (colorectal adenocarcinoma). These cytotoxic effects may be due to the presence of *trans*-caryophyllene and / or germacrene D in the VP and fractions. The cytotoxic activity of *trans*-caryophyllene on animal and human tumor cells has been previously described [75]. It is noteworthy that these fractions had little or no cytotoxic effect against the non-tumor cell line, HS5, suggesting selective cytotoxic activity for these fractions. This would be an important biological effect and could provide new rational basis for the design of new antitumor compounds [75].

Essential oils rich in chamazulene showed variable growth-inhibitory effects on human cancer cell lines with GI50 values ranging from 14.3 µg/mL on A375 (human malignant melanoma cell line) to 59.8 µg/mL on T98G (human glioblastoma cell line) [76] and strong antifungal properties against dermatophytes and opportunistic saprophytes [77]. Chamazulene has been reported to be an effective inhibitor of lipid peroxidation [76] with anti-inflammatory activity *in vivo* [78, 79]. However there are no reports on antiparasitic effects of chamazulene or dihydrochamazulene.

Trans-caryophyllene has been reported as antinociceptive [80], anxiolytic, antidepressant [81] and antimicrobial [82], among other effects. Essential oils rich in *trans*-caryophyllene and caryophyllene oxide from different plant species have been tested on *T. cruzi, T. vaginalis* [83 - 89] and other protozoan parasites such as *T. brucei*, *Plasmodium falciparum* and *L. infantum* [87, 90, 91] with different activity ranges alone or showing synergistic effects with the different EO components [88, 92, 93]. *Trans*-caryophyllene was effective against *T. cruzi* and *T. vaginalis,* but the activity of the domesticated *A. absinthium* oil fraction VLC1 was not fully explained by its content in *trans*-caryophyllene, suggesting synergistic effects [75].

CONCLUDING REMARKS

Two thujone free Spanish populations of *Artemisia absinthium* from Teruel and Sierra Nevada (Spain) belonging to the *cis*-epoxyocimene + chrysanthenyl acetate type and the *cis*-epoxyocimene type, have been submitted to a domestication process. The domestication resulted in lower chemical variation and higher biomass and essential oil yields, allowing for the registry of a new domesticated variety (® Candial). The main components of the new variety essential oil extracts were *cis*-epoxyocimene, Chrysanthenol, Chrysanthenyl acetate, *trans*-Caryophyllene and Linalool. (*5Z*)-2,6-dimethylocta-5,7-diene-2,3-diol was only present in the hydro-distilled HD extracts but not in the semi-industrial vapor-pressure extracts (VPs). The presence of 1.8-cineol and camphor correlated with the domestication level of *A. absinthium*. Supercritical fluid extraction, compared to traditional extraction techniques improved the yield of (*Z*)- epoxyocimene, chrysanthenol, (*5Z*)-2,6-dimethylocta-5,7-diene-2,3-diol and *trans*-chrysanthenyl acetate . The development of extraction processes with supercritical CO_2 allows for the enrichment of desired bioactive compounds in *A. absinthium* extracts. The supercritical extracts showed strong antifeedant and insecticidal effects against Spodoptera littoralis, indicating that for insect control purposes supercritical CO_2 extraction should be used.

The semi-industrial vapor-pressure extracts (VPs) showed strong antifungal effects. The antifungal effects of hydro distilled essential oils (HDs) were lower than the vapor pressure ones. Therefore, for antifungal applications the domesticated population submitted to vapor-pressure extraction should be used. (-)-cis-chrysanthenol has been identified as the main antifungal agent present in the VP extracts, followed by linalool. Therefore, (-)-*cis*-chrysanthenol should be used as the chemical indicator of the antifungal quality of VP extracts from *A. absinthium* (® Candial).

The semi-industrial vapor-pressure extracts (VPs) showed a high nematocidal activity against animal parasitic nematode *Trichinella spiralis* without being cytotoxic to macrophage cells. These oils showed activity on the protozoan parasites *Trypanosoma cruzi* and *Trichomonas vaginalis*. The antiprotozoal activity of the VP oil was related with the presence of *trans*-Caryophyllene and dihydrochamazulene. The VP oils also inhibited the growth of human tumour cell lines in a dose-dependent manner. *cis*-epoxyocimene + *cis*-chrysanthenol chemotype EOs from cultivated *A. absinthium* could be considered as potential candidates for development of new drugs against helminth and protozoan parasites of man and domestic animals.

CONFLICT OF INTEREST

The authors declare no conflict of interest, financial or otherwise.

ACKNOWLEDGEMENTS

This work has been supported by grant CTQ2015-64049-C3-1-R, (MINECO / FEDER).

REFERENCES

[1] Lachenmeier DW. Wormwood (*Artemisia absinthium* L.)--a curious plant with both neurotoxic and neuroprotective properties? J Ethnopharmacol 2010; 131(1): 224-7. [http://dx.doi.org/10.1016/j.jep.2010.05.062] [PMID: 20542104]

[2] Bora KS, Sharma A. Phytochemical and pharmacological potential of *Artemisia absinthium* Linn. and *Artemisia asiatica* Nakai: A Review. J Pharm Res 2010; 3: 325-8.

[3] Baykan Erel S, Reznicek G, Şenol SG, Karabay Yavaşogulu NÜ, Konyalioglu S, Zeybek AU. Antimicrobial and antioxidant properties of *Artemisia* L. species from western Anatolia. Turk J Biol 2012; 36: 75-84.

[4] Gandomi Nasrabadi H, Abbaszadeh S, Tayyar Hashtjin N, Yamrali I. Study of chemical composition of essential oil of afsantine (*Artemisia absinthium*) and inhibitory effects of the essential oil and its aqueous and alcoholic extracts on some food borne bacterial pathogens. Faslnamah-i Giyahan-i Daruyi 2012; 11: 120-7.

[5] Juteau F, Jerkovic I, Masotti V, *et al.* Composition and antimicrobial activity of the essential oil of *Artemisia absinthium* from Croatia and France. Planta Med 2003; 69(2): 158-61. [http://dx.doi.org/10.1055/s-2003-37714] [PMID: 12624823]

[6] Tariku Y, Hymete A, Hailu A, Rohloff J. *in vitro* evaluation of antileishmanial activity and toxicity of essential oils of *Artemisia absinthium* and *Echinops kebericho*. Chem Biodivers 2011; 8(4): 614-23. [http://dx.doi.org/10.1002/cbdv.201000331] [PMID: 21480507]

[7] Chiasson H, Bélanger A, Bostanian N, Vincent C, Poliquin A. Acaricidal properties of *Artemisia absinthium* and *Tanacetum vulgare* (Asteraceae) essential oils obtained by three methods of extraction. J Econ Entomol 2001; 94(1): 167-71. [http://dx.doi.org/10.1603/0022-0493-94.1.167] [PMID: 11233109]

[8] Kaul VK, Nigam SS, Banerjee AK. Insecticidal activity of some essential oils. Indian J Pharm 1978; 40: 22.

[9] Kordali S, Aslan I, Çalmaşur O, Cakir A. Toxicity of essential oils isolated from three *Artemisia* species and some of their major components to granary weevil, *Sitophilus granarius* (L.) (Coleoptera: Curculionidae). Ind Crops Prod 2006; 23: 162-70. [http://dx.doi.org/10.1016/j.indcrop.2005.05.005]

[10] Umpiérrez ML, Lagreca ME, Cabrera R, Grille G, Rossini C. Essential oils from Asteraceae as potential biocontrol tools for tomato pests and diseases. Phytochem Rev 2012; 11: 339-50. [http://dx.doi.org/10.1007/s11101-012-9253-5]

[11] Erichsen-Brown C. Use of Plants for the Past 500 Years. Ontario, Canada: Aurora 1979.

[12] Jaenson TG, Pålsson K, Borg-Karlson AK. Evaluation of extracts and oils of tick-repellent plants from Sweden. Med Vet Entomol 2005; 19(4): 345-52. [http://dx.doi.org/10.1111/j.1365-2915.2005.00578.x] [PMID: 16336298]

[13] Carnat AP, Madesclaire M, Chavignon O, Lamaison JI. cis-chrysanthenol, a main component in essential oil of *Artemisia absinthium* L. growing in Auvergne (Massif Central), France. J Essent Oil Res 1992; 4: 487-90. [http://dx.doi.org/10.1080/10412905.1992.9698115]

[14] Chialva F, Liddle AP, Doglia G. Chemotaxonomy of wormwood (*Artemisia absinthium* L.) I. Composition of the essential oil of several chemotypes. Z Lebensm Unters Forsch 1983; 176: 363-6.

[15] Geszprych A, Przybył J, Kuczerenko A, Węglarz Z. Diversity of Wormwood (*Artemisia absinthium* L.) Growing Wild in Poland in Respect of the Content and Composition of Essential Oil and Phenolic Compounds. Acta Hortic 2011; 123-9. [http://dx.doi.org/10.17660/ActaHortic.2011.925.16]

[16] Judzentiene A, Budiene J, Gircyte R, Masotti V, Laffont-Schwob I. Toxic Activity and Chemical Composition of Lithuanian Wormwood (*Artemisia absinthium* L.) Essential Oils. Rec Nat Prod 2012; 6: 180-3.

[17] Karp F, Croteau R. Evidence that sabinene is an essential precursor of C(3)-oxygenated thujane monoterpenes. Arch Biochem Biophys 1982; 216(2): 616-24. [http://dx.doi.org/10.1016/0003-9861(82)90251-X] [PMID: 7051980]

[18] Lopes-Lutz D, Alviano DS, Alviano CS, Kolodziejczyk PP. Screening of chemical composition, antimicrobial and antioxidant activities of *Artemisia* essential oils. Phytochemistry 2008; 69(8): 1732-8. [http://dx.doi.org/10.1016/j.phytochem.2008.02.014] [PMID: 18417176]

[19] Sharopov FS, Sulaimonova VA, Setzer WN. Composition of the Essential oil of *Artemisia absinthium* from Tajikistan. Rec Nat Prod 2012; 6: 127-34.

[20] Tehrani MS, Azar PA, Hosain SW, Khalilzadeh MA, Zanousi MB. Composition of essential oil of *Artemisia absinthium* by three different extrac-tion methods: Hydrodistillation, solvent-free microwave extraction & headspace solid-phase microextraction. Asian J Chem 2012; 24: 5371-6.

[21] Chialva F, Doglia G, Gabri G, Aime S, Milone L. Isolamento ed identificazione del cis e trans-epossiocimene nell olio essenziale de *Artemisia absinthium* Linnaeus italiana. Riv Ital EPPOS 1976; 10: 522-35.

[22] Pino JA, Rosado A, Fuentes V. Chemical composition of the essential oil of *Artemisia absinthium* L. from Cuba. J Essent Oil Res 1997; 9: 87-9. [http://dx.doi.org/10.1080/10412905.1997.9700720]

[23] Leung AY. Encyclopedia of Common Natural Ingredients used in Food, Drugs, and Cosmetics. New York: Wiley 1980.

[24] Martín L, Julio LF, Burillo J, Sanz J, Mainar AM, González-Coloma A. Comparative chemistry and insect antifeedant action of traditional (Clevenger and Soxhlet) and supercritical extracts (CO2) of two cultivated wormwood (*Artemisia absinthium* L.) populations. Ind Crops Prod 2011; 34: 1615-21. a

[http://dx.doi.org/10.1016/j.indcrop.2011.06.006]

[25] Danijela V, Sanja ĆZ, Muamer D. Milka. M. Essential oil composition and antioxidant activity of four Asteraceae species from Bosnia. J Essent Oil Res 2016; 28(5): 445-57. [http://dx.doi.org/10.1080/10412905.2016.1150216]

[26] Ariño A, Arberas I, Renobales G, Arriaga S, Dominguez JB. Essential oil of *Artemisia absinthium* L. from the Spanish pyrenees. J Essent Oil Res 1999; 11: 182-4. [http://dx.doi.org/10.1080/10412905.1999.9701105]

[27] Bailen M, Julio LF, Diaz CE, Sanz J, Martínez-Díaz RA, Cabrera R, *et al.* Chemical composition and biological effects of essential oils from *Artemisia absinthium* L. cultivated under different environmental conditions. Ind Crops Prod 2013; 49: 102-7. [http://dx.doi.org/10.1016/j.indcrop.2013.04.055]

[28] Burillo J. Cultivo experimental de ajenjo Artemisia absinthium L. como potencial insecticida de origen natural.Insecticidas y Repelentes De Origen Natural. Centro de Investigación y Tecnología Agroalimentaria Zaragoza 2009; pp. 19-30.

[29] Gonzalez-Coloma A, Bailen M, Diaz CE, Fraga BM, Martínez-Díaz R, Zuñiga GE, *et al.* Major components of Spanish cultivated *Artemisia absinthium* populations: Antifeedant, antiparasitic, and antioxidant effects. Ind Crops Prod 2012; 37: 401-7. [http://dx.doi.org/10.1016/j.indcrop.2011.12.025]

[30] Martín L, González-Coloma A, Burillo J, Palavra AM, Urieta JS, Mainar AM. Microcalorimetric determination of the activity of supercritical extracts of wormwood (*Artemisia absinthium* L.) over Spodoptera littoralis. J Therm Anal Calorim 2012; 111: 1837-44. [http://dx.doi.org/10.1007/s10973-012-2661-3]

[31] Sainz P, Sanz J, Burillo J, González-Coloma A, Bailén M. R.A. M-D. Essential oils for the control of reduviid insects. Phytochem Rev 2012; 11: 361-9. [http://dx.doi.org/10.1007/s11101-012-9261-5]

[32] Julio LF, Burillo J, Giménez C, Cabrera R, Díaz CE, Sanz J, *et al.* Chemical and biocidal characterization of two cultivated *Artemisia absinthium* populations with different domestication levels. Ind Crops Prod 2015; 76: 787-92. [http://dx.doi.org/10.1016/j.indcrop.2015.07.041]

[33] Osorio C, Duque C, Suárez M, Salamanca LE, Urueña F. Free, glycosidically bound, and phosphate bound flavor constituents of badea (*Passiflora quadrangularis*) fruit pulp. J Sep Sci 2002; 25: 147-54. [http://dx.doi.org/10.1002/1615-9314(20020201)25:3<147::AID-JSSC147>3.0.CO;2-G]

[34] Core RJ, Henning JA, Gardea-Torresdey J, Mostafavi R. Quantitative comparison of volatile compounds among seven*Medicago* spp. accessions. J Chem Ecol 1996; 22(9): 1621-7. [http://dx.doi.org/10.1007/BF02272402] [PMID: 24226475]

[35] Tsankova E, Bohlmann F. A monoterpene from Aster bakeranus. Phytochemistry 1983; 22: 1285-6. [http://dx.doi.org/10.1016/0031-9422(83)80244-1]

[36] Gholami M, Azizi A, Salehi P. Variations in essential oil components in cultivated and regenrated *Artemisia absinthium* L. Asian J Chem 2005; 17: 2229-32.

[37] Blagojević P, Radulović N, Palić R, Stojanović G. Chemical composition of the essential oils of serbian wild-growing *Artemisia absinthium* and *Artemisia vulgaris*. J Agric Food Chem 2006; 54(13): 4780-9. [http://dx.doi.org/10.1021/jf060123o] [PMID: 16787028]

[38] Pourmortazavi SM, Hajimirsadeghi SS. Supercritical fluid extraction in plant essential and volatile oil analysis. J Chromatogr A 2007; 1163(1-2): 2-24. [http://dx.doi.org/10.1016/j.chroma.2007.06.021] [PMID: 17624357]

[39] Stahl E, Gerard D. Hochdruck-Extraktion von Naturstoffen mit überkritischen und verflüssigten Gasen. Zeitschrift Für Lebensmitteluntersuchung Und -Forschung. 1983; A 176: pp. 1-4.

[40] Ramasubramaniaraja R. M NB. Antihelminthic studies and medicinal herbs – an overview. Int J Pharm Sci Rev Res 2010; 5: 39-47.

[41] Wiley/NBS Registry of Mass Spectral Data. Mc Lafferty FW & Stauffe DB NY 1989.

[42] D'Addabbo T, Carbonara T, Argentieri MP, Radicci V, Leonetti P, Villanova L, *et al.* Nematicidal potential of *Artemisia annua* and its main metabolites. Eur J Plant Pathol 2013; 137(2): 295-304. [http://dx.doi.org/10.1007/s10658-013-0240-5]

[43] Squires JM, Ferreira JF, Lindsay DS, Zajac AM. Effects of artemisinin and Artemisia extracts on *Haemonchus contortus* in gerbils (*Meriones unguiculatus*). Vet Parasitol 2011; 175(1-2): 103-8. [http://dx.doi.org/10.1016/j.vetpar.2010.09.011] [PMID: 20943323]

[44] Tariq KA, Chishti MZ, Ahmad F, Shawl AS. Anthelmintic activity of extracts of *Artemisia absinthium* against ovine nematodes. Vet Parasitol 2009; 160(1-2): 83-8. [http://dx.doi.org/10.1016/j.vetpar.2008.10.084] [PMID: 19070963]

[45] Bird DM, Bird AF. Plant parasitic nematodes.Parasitic Nematodes: Molecular Biology CABI. Wallingford: Biochemistry and Immunology 2001; pp. 139-66. [http://dx.doi.org/10.1079/9780851994239.0139]

[46] Pozio E. The broad spectrum of *Trichinella* hosts: from cold- to warm-blooded animals. Vet Parasitol 2005; 132(1-2): 3-11. [http://dx.doi.org/10.1016/j.vetpar.2005.05.024] [PMID: 15970384]

[47] de Aquino Mesquita M, E Silva Júnior JB, Panassol AM, *et al.* Anthelmintic activity of Eucalyptus staigeriana encapsulated oil on sheep gastrointestinal nematodes. Parasitol Res 2013; 112(9): 3161-5. [http://dx.doi.org/10.1007/s00436-013-3492-2] [PMID: 23783400]

[48] García JJ, Bolás F, Torrado JJ. Bioavailability and efficacy characteristics of two different oral liquid formulations of albendazole. Int J Pharm 2003; 250(2): 351-8. [http://dx.doi.org/10.1016/S0378-5173(02)00559-8] [PMID: 12527162]

[49] Prichard RK. Markers for benzimidazole resistance in human parasitic nematodes? Parasitology 2007; 134(Pt 8): 1087-92. [http://dx.doi.org/10.1017/S003118200700008X] [PMID: 17608968]

[50] Rassi A Jr, Rassi A, Marin-Neto JA. Chagas disease. Lancet 2010; 375(9723): 1388-402. [http://dx.doi.org/10.1016/S0140-6736(10)60061-X] [PMID: 20399979]

[51] WHO-World Health Organization. Global Incidence and Prevalence of Selected Curable Sexually Transmitted Infections-2008 2012.apps.who.int/iris/bitstream/ 0665/75181/1/9789241503839_ eng.pdf?ua=1

[52] Cudmore SL, Delgaty KL, Hayward-McClelland SF, Petrin DP, Garber GE. Treatment of infections caused by metronidazole-resistant *Trichomonas vaginalis.* Clin Microbiol Rev 2004; 17(4): 783-93. [http://dx.doi.org/10.1128/CMR.17.4.783-793.2004] [PMID: 15489348]

[53] Rodilla JM, Tinoco MT, Morais JC, Gimenez C, Cabrera R, Martín-Benito D, *et al. Laurus novocanariensis* essential oil: Seasonal variation and valorization. Biochem Syst Ecol 2008; 36: 167-76. [http://dx.doi.org/10.1016/j.bse.2007.09.001]

[54] Agnaniet H, Agrebi A, Bikanga R, *et al.* Essential oil of *Plectranthus tenuicaulis* leaves from Gabon, source of (R),(E)-6,7-epoxyocimene. An unusual chemical composition within the genus *Plectranthus.* Nat Prod Commun 2011; 6(3): 409-16. [PMID: 21485283]

[55] Pragadheesh VS, Saroj A, Yadav A, Chanotiya CS, Alam M, Samad A. Chemical characterization and antifungal activity of *Cinnamomum camphora* essential oil. Ind Crops Prod 2013; 49: 628-33. [http://dx.doi.org/10.1016/j.indcrop.2013.06.023]

[56] Shimada T, Endo T, Fujii H, Rodriguez A, Pena L, Omura M. Characterization of three linalool

synthase genes from *Citrus unshiu* Marc. and analysis of linalool-mediated resistance against *Xanthomonas citri* subsp. citri and *Penicilium italicum* in citrus leaves and fruits. Plant science : an international journal of experimental plant biology 2014; 229: 154-66.

[57] Tanaka K, Taniguchi S, Tamaoki D, Yoshitomi K, Akimitsu K, Gomi K. Multiple roles of plant volatiles in jasmonate-induced defense response in rice. Plant Signal Behav 2014; 9: e29247. [http://dx.doi.org/10.4161/psb.29247]

[58] García-Rodríguez JJ, Andrés MF, Ibañez-Escribano A, *et al.* Selective nematocidal effects of essential oils from two cultivated *Artemisia absinthium* populations. Z Natforsch C J Biosci 2015; 70(9-10): 275-80. [http://dx.doi.org/10.1515/znc-2015-0109] [PMID: 26444350]

[59] Giarratana F, Muscolino D, Beninati C, Giuffrida A, Panebianco A. Activity of *Thymus vulgaris* essential oil against *Anisakis* larvae. Exp Parasitol 2014; 142: 7-10. [http://dx.doi.org/10.1016/j.exppara.2014.03.028] [PMID: 24721259]

[60] Macedo IT, Bevilaqua CM, de Oliveira LM, *et al.* Anthelmintic effect of *Eucalyptus staigeriana* essential oil against goat gastrointestinal nematodes. Vet Parasitol 2010; 173(1-2): 93-8. [http://dx.doi.org/10.1016/j.vetpar.2010.06.004] [PMID: 20609526]

[61] Macedo IT, de Oliveira LM, Camurça-Vasconcelos AL, *et al. in vitro* effects of *Coriandrum sativum, Tagetes minuta, Alpinia zerumbet* and *Lantana camara* essential oils on Haemonchus contortus. Rev Bras Parasitol Vet 2013; 22(4): 463-9. [http://dx.doi.org/10.1590/S1984-29612013000400004] [PMID: 24473869]

[62] Zhu L, Dai J, Yang L, Qiu J. Anthelmintic activity of *Arisaema franchetianum* and *Arisaema lobatum* essential oils against *Haemonchus contortus.* J Ethnopharmacol 2013; 148(1): 311-6. [http://dx.doi.org/10.1016/j.jep.2013.04.034] [PMID: 23624115]

[63] Andres MF, Gonzalez-Coloma A, Sanz J, Burillo J, Sainz P. Nematicidal activity of essential oils: a review. Phytochem Rev 2012; 11: 371-90. [http://dx.doi.org/10.1007/s11101-012-9263-3]

[64] Bai PH, Bai CQ, Liu QZ, Du SS, Liu ZL. Nematicidal activity of the essential oil of Rhododendron anthopogonoides aerial parts and its constituent compounds against Meloidogyne incognita. Z Natforsch C J Biosci 2013; 68(7-8): 307-12. [http://dx.doi.org/10.5560/ZNC.2013.68c0307] [PMID: 24066516]

[65] Li W, Sun YN, Yan XT, *et al.* Isolation of nematicidal triterpenoid saponins from *Pulsatilla koreana* root and their activities against *Meloidogyne incognita.* Molecules 2013; 18(5): 5306-16. [http://dx.doi.org/10.3390/molecules18055306] [PMID: 23698044]

[66] Mattei D, Dias-Arieira CR, Biela F, Roldi M, Benetoli da Silva TR, Rampim L, *et al.* Essential oil of Rosmarinus officinalis in the control of *Meloidogyne javanica* and *Pratylenchus brachyurus* in soybean. Biosci J 2014; 30: 469-76.

[67] Jasmer DP, Goverse A, Smant G. Parasitic nematode interactions with mammals and plants. Annu Rev Phytopathol 2003; 41: 245-70. [http://dx.doi.org/10.1146/annurev.phyto.41.052102.104023] [PMID: 14527330]

[68] Ferreira JF, Peaden P, Keiser J. *in vitro* trematocidal effects of crude alcoholic extracts of *Artemisia annua, A. absinthium, Asimina triloba,* and *Fumaria officinalis*: trematocidal plant alcoholic extracts. Parasitol Res 2011; 109(6): 1585-92. [http://dx.doi.org/10.1007/s00436-011-2418-0] [PMID: 21562762]

[69] Caner A, Döşkaya M, Değirmenci A, *et al.* Comparison of the effects of *Artemisia vulgaris* and *Artemisia absinthium* growing in western Anatolia against trichinellosis (*Trichinella spiralis*) in rats. Exp Parasitol 2008; 119(1): 173-9. [http://dx.doi.org/10.1016/j.exppara.2008.01.012] [PMID: 18325496]

[70] Barbosa P, Lima AS, Vieira P, *et al.* Nematicidal activity of essential oils and volatiles derived from

Portuguese aromatic flora against the pinewood nematode, *Bursaphelenchus xylophilus*. J Nematol 2010; 42(1): 8-16.
[PMID: 22736831]

[71] Zouhar M, Douda O, Lhotsky D, Pavela R. Effect of plant essential oils on mortality of the stem nematode (*Ditylenchus dipsaci*). Plant Prot Sci 2009; 45: 66-73.

[72] López de Mendoza ME, Abrantes IM, Rowe J. RH C. Immunolocalisation in plants of secretions form parasitic stages of *Meloidogyne incognita* and *M. hispanica*. Int J Nematol 2002; 12: 149-54.

[73] López de Mendoza ME, Curtis RH, Gowen S. Identification and characterization of excreted-secreted products and surface coat antigens of animal and plant-parasitic nematodes. Parasitology 1999; 118(Pt 4): 397-405.
[http://dx.doi.org/10.1017/S0031182098003941] [PMID: 10340331]

[74] López-Arellano ME. Immunomolocalization of *Trichinella spiralis* L1 surece and excreted/secreted antigens *in situ*. Int J Nematol 2002; 12: 55-8.

[75] Martínez-Díaz RA, Ibáñez-Escribano A, Burillo J, *et al.* Trypanocidal, trichomonacidal and cytotoxic components of cultivated *Artemisia absinthium* Linnaeus (Asteraceae) essential oil. Mem Inst Oswaldo Cruz 2015; 110(5): 693-9.
[http://dx.doi.org/10.1590/0074-02760140129] [PMID: 26107187]

[76] El Hadri A, Del Río MAG, Sanz J, González-Coloma A, Idaomar M, Ozonas BR, *et al.* Cytotoxic activity of α-humulene and trans-caryophyllene from *Salvia officinalis* in animal and human tumor cells. An R Acad Nac Farm 2010; 76: 343-56.

[77] Jamalian A, Shams-Ghahfarokhi M, Jaimand K, Pashootan N, Amani A, Razzaghi-Abyaneh M. Chemical composition and antifungal activity of *Matricaria recutita* flower essential oil against medically important dermatophytes and soil-borne pathogens. J Mycol Med 2012; 22(4): 308-15.
[http://dx.doi.org/10.1016/j.mycmed.2012.09.003] [PMID: 23518164]

[78] Ornano L, Venditti A, Ballero M, *et al.* Chemopreventive and antioxidant activity of the chamazulene-rich essential oil obtained from *Artemisia arborescens* L. growing on the Isle of La Maddalena, Sardinia, Italy. Chem Biodivers 2013; 10(8): 1464-74.
[http://dx.doi.org/10.1002/cbdv.201200435] [PMID: 23939794]

[79] Rekka EA, Kourounakis AP, Kourounakis PN. Investigation of the effect of chamazulene on lipid peroxidation and free radical processes. Res Commun Mol Pathol Pharmacol 1996; 92(3): 361-4.
[PMID: 8827832]

[80] Safayhi H, Sabieraj J, Sailer ER, Ammon HP. Chamazulene: an antioxidant-type inhibitor of leukotriene B4 formation. Planta Med 1994; 60(5): 410-3.
[http://dx.doi.org/10.1055/s-2006-959520] [PMID: 7997466]

[81] Katsuyama S, Mizoguchi H, Kuwahata H, *et al.* Involvement of peripheral cannabinoid and opioid receptors in β-caryophyllene-induced antinociception. Eur J Pain 2013; 17(5): 664-75.
[http://dx.doi.org/10.1002/j.1532-2149.2012.00242.x] [PMID: 23138934]

[82] Bahi A, Al Mansouri S, Al Memari E, Al Ameri M, Nurulain SM, Ojha S. β-Caryophyllene, a CB2 receptor agonist produces multiple behavioral changes relevant to anxiety and depression in mice. Physiol Behav 2014; 135: 119-24.
[http://dx.doi.org/10.1016/j.physbeh.2014.06.003] [PMID: 24930711]

[83] Cheikh-Ali Z, Adiko M, Bouttier S, *et al.* Composition, and antimicrobial and remarkable antiprotozoal activities of the essential oil of rhizomes of *Aframomum sceptrum* K. Schum. (Zingiberaceae). Chem Biodivers 2011; 8(4): 658-67.
[http://dx.doi.org/10.1002/cbdv.201000216] [PMID: 21480511]

[84] Costa EV, Dutra LM, Salvador MJ, Ribeiro LH, Gadelha FR, de Carvalho JE. Chemical composition of the essential oils of Annona pickelii and *Annona salzmannii* (Annonaceae), and their antitumour and trypanocidal activities. Nat Prod Res 2013; 27(11): 997-1001.

[http://dx.doi.org/10.1080/14786419.2012.686913] [PMID: 22583044]

[85] da Silva TB, Menezes LR, Sampaio MF, *et al.* Chemical composition and anti-*Trypanosoma cruzi* activity of essential oils obtained from leaves of *Xylopia frutescens* and *X. laevigata* (Annonaceae). Nat Prod Commun 2013; 8(3): 403-6.
[PMID: 23678822]

[86] Goren AC, Piozzi F, Akcicek E, Kılıc T. Essential oil composition of twenty-two *Stachys* species (mountain tea) and their biological activities. Phytochem Lett 2011; 4: 448-53.
[http://dx.doi.org/10.1016/j.phytol.2011.04.013]

[87] Monzote L, García M, Pastor J, *et al.* Essential oil from *Chenopodium ambrosioides* and main components: activity against Leishmania, their mitochondria and other microorganisms. Exp Parasitol 2014; 136: 20-6.
[http://dx.doi.org/10.1016/j.exppara.2013.10.007] [PMID: 24184772]

[88] Polanco-Hernández G, Escalante-Erosa F, García-Sosa K, *et al.* Metabolites from the leaf extract of *Serjania yucatanensis* with trypanocidal activity against *Trypanosoma cruzi.* Parasitol Res 2012; 111(1): 451-5.
[http://dx.doi.org/10.1007/s00436-012-2861-6] [PMID: 22371270]

[89] Vermani K, Garg S. Herbal medicines for sexually transmitted diseases and AIDS. J Ethnopharmacol 2002; 80(1): 49-66.
[http://dx.doi.org/10.1016/S0378-8741(02)00009-0] [PMID: 11891087]

[90] Gachet MS, Kunert O, Kaiser M, *et al.* Antiparasitic compounds from *Cupania cinerea* with activities against *Plasmodium falciparum* and *Trypanosoma brucei rhodesiense.* J Nat Prod 2011; 74(4): 559-66.
[http://dx.doi.org/10.1021/np100415m] [PMID: 21438586]

[91] Nibret E, Wink M. Trypanocidal and antileukaemic effects of the essential oils of *Hagenia abyssinica, Leonotis ocymifolia, Moringa stenopetala*, and their main individual constituents. Phytomedicine 2010; 17(12): 911-20.
[http://dx.doi.org/10.1016/j.phymed.2010.02.009] [PMID: 20359874]

[92] Polanco-Hernandez G, Escalante-Erosa F, Garcia-Sosa K, Rosado ME, Guzman-Marin E, Acosta-Viana KY, *et al.* Synergistic Effect of Lupenone and Caryophyllene Oxide against Trypanosoma cruzi. Evid Based Complement Alternat Med 2013; 2013: 435398.

[93] Setzer W, Stokes S, Bansal A, Haber W, Caffrey C, Hansell E, *et al.* Chemical composition and cruzain inhibitory activity of *Croton draco* bark essential oil from Monteverde, Costa Rica. Nat Prod Commun 2007; 2: 685-9.

SUBJECT INDEX

A

M. Paz Arraiza, A. González-Coloma, J. Burillo and C. Calderón-Guerrero (Eds.)

B

C

F

G

H

I

J

L

M

Q

R

S

T

U

V

W

Z

www.ingramcontent.com/pod-product-compliance
Lightning Source LLC
LaVergne TN
LVHW070950210926
848482LV00002B/170